EDUCATION IN A COMPETITIVE AND GLOBALIZING WORLD

FROM PCK TO TPACK

RESEARCH AND DEVELOPMENT

EDUCATION IN A COMPETITIVE AND GLOBALIZING WORLD

Additional books in this series can be found on Nova's website under the Series tab.

Additional e-books in this series can be found on Nova's website under the e-book tab.

EDUCATION IN A COMPETITIVE AND GLOBALIZING WORLD

FROM PCK TO TPACK

RESEARCH AND DEVELOPMENT

JANG SYH-JONG

Nova Science Publishers, Inc.
New York

Library of Congress Cataloging-in-Publication Data

From PCK to TPACK : research and development / Jang Syh-Jong (Graduate School of Education, Chung-Yuan Christian University, Chung-Li, Taiwan).
pages cm
Includes bibliographical references and index.
ISBN 978-1-62081-146-7 (hardcover)
1. Science--Study and teaching--Taiwan. 2. Science--Study and teaching (Middle school)--Taiwan. 3. Science teachers--Training of--Taiwan. 4. Pedagogical content knowledge--Taiwan. I. Jang, Syh-Jong.
Q183.4.T28F76 2012
507.1'051249--dc23

2012005428

Published by Nova Science Publishers, Inc. † New York

CONTENTS

PREFACE

This particular year is worthwhile to remember for my sabbatical break. I dream and intend to write a book related to PCK and TPACK research not only for elementary and middle school teachers, but also for university instructors. In recent years, there has been growing interest associated with the notion of the professional development teachers; however, few books have been published for this particular issue. The Ministry of Education (Taiwan) provided a lot of financial support to all levels of schools in order to enhance teachers' professional development. However, many novice college teachers with doctoral degrees and a certain level of their subject matter knowledge could not teach effectively. The teachers' pedagogical content knowledge (PCK) is the key to advancing professional growth with all levels of teachers. On the other hand, the current issue at hand is how to integrate technology into a teacher's PCK. In contrast to the simple view of technology treated merely as a tool, school teachers in the 21st century need to be equipped with professional teacher knowledge, referred to as Technological Pedagogical and Content Knowledge (TPACK). This implies that teachers have the ability to apply technology into their pedagogical strategies and content representations for teaching specific topics to promote their students' learning efficiency. Furthermore, the TPACK framework not only expresses the importance of technological integration, but introduces the relationships and complexities between all three basic components of knowledge (technology, pedagogy, and content).

This book is divided into two parts. In the first part, there are four studies related to PCK research and development (chapters 1-4). Chapter one followed the model of Magnusson, Krajcik and Borko (1999), and an instrument was developed with satisfactory content and construct validity in this study. We found the didactic orientation was the most commonly used method of science teaching in Taiwanese middle schools. It was found the didactic orientation was negatively related to KIS, and KASL was significantly correlated to activity-driven orientation. The results indicated that KSC was found to be significantly different by gender, and science teachers' overall PCK and three components (i.e., KSC, KSUS and KASL) were significantly different according to teaching experience. Chapter two aims to develop an instrument that could be employed to evaluate university students' perceptions of teachers' Pedagogical Content knowledge. According to the pilot test results, the four categories finally constructed were Subject Matter Knowledge (SMK), Instructional Representation and Strategies (IRS), Instructional Objects and Context (IOC), and Knowledge of Students' Understanding (KSU). In the main study, seven items were generated under each category, making a total of 28 items in the questionnaire, and the instrument was administered to 172 college education students. The analysis results indicated that the instrument referring to teachers' knowledge shows satisfactory validity and reliability.

Chapter three evaluates a reflective model and how it affects university students' perceptions of teachers' PCK development. A developed survey and reflective model were the tools used for six novice teachers to facilitate reflection and exchange ideas among teachers. This study adopted a case study approach and conducted both quantitative and qualitative analyses. We also compared, in detail, the students' perception of six novice teachers' PCK performance, individually in four dimensions. In particular, this study clearly describes each teacher's teaching performance before and after the reflection. Chapter four described 26 university instructors who participated in all three workshops and whose students receive both the pre- and post-tests. PCK workshops were held for these university instructors to facilitate their self-reflection and the exchange of ideas among teachers. Data was collected, summarized and coded by surveys, interviews and PCK workshops. The results showed that more than half of the university instructors showed improvement in their PCK. Students considered that university instructors were strongest in Subject Matter Knowledge (SMK), and weaker in Instructional Representation and Strategies (IRS) and Knowledge of Students' Understanding (KSU). The workshops provided instructors with opportunities for learning from each other and for reflecting on their teaching methods.

In the second part, there are four studies related to TPACK research, development and application (chapters 5-8). Chapter five aims to assess the university students' perception of the case instructor's TPACK development. A physics instructor and his class, a total of 70 students, participated in this study. This study used a mixed method design, incorporating both quantitative and qualitative techniques. A developed questionnaire was distributed to the students at the middle and end of the semester, respectively. The study adopted constant comparison and the triangulation method related to the instructor's interviews, students' opinions and quantitative data to explain the instructor's TPACK change. This study provided valuable insight and details related to the development of the cvase teacher's TPACK and provided program feedback on effective approaches in encouraging this development. Chapter six describes that Interactive Whiteboard (IWB) technology and the TPACK of teachers are the important issues. However, it would appear the means for training science teachers to use the IWB and assessing the TPACK of science teachers have yet to be fully investigated. This study provides a theoretical framework for the IWB-TPACK model and develops evaluative criteria from an integrative view to assess the case science teacher's TPACK. The case teacher and her 32 seventh grade students participated in this study. Two instances of classroom teaching observations were conducted. Results showed that the case teacher's TPACK improved on the whole, especially in technology-related knowledge, and appropriate use of animations and videos as part of the IWB. Furthermore, the observation produced a teaching reflection on the case teacher, who used the IWB to enhance teacher-student interaction and evaluate student knowledge.

Chapter seven examines how peer coaching affects the TPACK development of science teachers and a student's learning outcomes. A peer coaching-based model for TPACK was developed. Three certified science teachers and three classes of eighth graders were assigned to the experimental group, and three science teachers and three similar classes were assigned to the control group. The main data included students' scores, written assignments, reflective journals and interviews. The results showed that students in the experimental group scored higher than those in the control group, and the two groups had a significant difference in respect to the students' achievement. The experimental science teachers were clearly aware of students' prior conceptions of the subject matter and their learning difficulties. They also learned to implement technology and multiple teaching methods in order to integrate the subject-matter knowledge. Chapter eight describes a collaborative action research employed within the progresses of planning, implementation, observation, reflection, and evaluation. The full IWB implementation took approximately one semester that consisted of two action circles. The participants were five elementary teachers from the same school, forming a peer community. Data was primarily collected by means of teaching reflection notes, in-class videotape, a peer community meeting and TPCK evaluation questionnaires. Research that has empirically documented the mathematics teachers' TPACK with the specific content knowledge is scant. This study showed the developing of the case teacher's TPACK process through two, circle action research. Along with the circular feedbacks, the results of the study also uncovered the limitations of technology. More specifically, when IWB was overly emphasized, the transformation of teaching content was restricted.

In conclusion, despite the PCK and TPACK research and development of elementary and middle school teachers (chapter 1, 6, 7, 8), the major contribution of this book includes the construction of theoretical bases and the development of an instrument for assessing university students' perceptions of their instructors' PCK and TPACK (chapter 2, 3, 4, 5). The use of the survey was to understand the overall teaching performance of the instructors and provide them with teaching reflection. Compared with traditional semester-end evaluations, which could only collect rather few opinions and postpone improvments in teaching the same subject, this instrument design allows to collect students' many opinions through open-ended questions and provides a diagnostic function to allow the instructors to find out their changes or differences after a period of teaching, and make reflective thinking as well as timely modifications.

This book can be published through many thanks and collaborations. I appreciate my wife, Phoebe, for her prayers and caring for my daughters. I also thank my assistants for typing and translating, and the NOVA publisher's support.

Syh-Jong, Jang
Professor
Graduate School of Education
Chung-Yuan Christian University
Chung-Li, Taiwan, R.O.C.
February 1, 2012

Chapter 1

DEVELOPING AN INSTRUMENT TO EXAMINE SCIENCE TEACHERS' PCK AND SCIENCE TEACHING ORIENTATIONS

ABSTRACT

Although there have been numerous pedagogical content knowledge (PCK) studies in science education, little attention has been paid to the link between PCK components and different types of orientations to science teaching (OST). Following the model of Magnusson, Krajcik and Borko (1999), an instrument was developed with satisfactory content and construct validity in this study. The purpose of this study was to examine the relationship that exists between all orientations to science teaching and PCK components, and explore whether the PCK of middle science teachers differs significantly according gender and teaching experience in Taiwan. We found the highest mean score, regarding the PCK of science teachers, was associated with the Knowledge of Science Curricula (KSC), whereas the lowest mean score was associated with the Knowledge of Assessment of Scientific Literacy (KASL). The results indicated that KSC was found to be significantly different by gender, and science teachers' overall PCK and three components (i.e., KSC, KSUS and KASL) were significantly different according to teaching experiences. On the other hand, the didactic orientation was the most commonly used method of science teaching in Taiwanese middle schools. According to the Pearson correlation analysis, the didactic orientation was found to be negatively related to KIS. We also found KASL was significantly correlated to an activity-driven orientation. Research implications of this study are provided, along with suggestions for future research.

Keywords: Pedagogical content knowledge; Instrument; Orientations to science teaching; Middle teachers

INTRODUCTION

The pedagogical content knowledge (PCK) of science teachers is currently an important issue for teacher education in Taiwan, and most of the emphasis has been on the development and research of PCK for science teachers (Abell, 2008; De Jong, Van Driel & Verloop, 2005; Gess-Newsome & Lederman, 1993; Grossman, 1990; Loughran, Mulhall & Berry, 2004;

Nilsson, 2008; Van Driel, De Jong, & Verloop, 2002; Van Driel, Verloop & de Vos, 1998). Due to the implementation of a nine-year integrated curriculum for primary and middle schools in Taiwan in 2004, research on the PCK of pre-service and in-service science teachers is an urgent matter. This newly revised curriculum, running from first through ninth grades, provides a continuous curriculum that integrates traditionally separate courses into several learning fields. The curriculum for each subject has been integrated into a larger field of interdisciplinary courses; for example, the previously separate courses of Biology, Physics, Chemistry, and Earth Science in middle school (seventh to ninth grade) were integrated under the single subject of Natural Science (Jang, 2006). This had a great impact on current middle science teachers of Taiwan, who needed to enhance their PCK and equip themselves with the abilities to integrate and design a curriculum, as well as aim for changes to develop more innovative teaching.

Science teachers' PCK has been one of the focal points in science teacher education, not only because it facilitates science teachers' teaching innovations (Van Dijk & Kattmann, 2007; Jang, 2009), but also because PCK could be one of the key factors that fosters student learning (Hanuscin, Lee, & Akerson, 2011; Park & Oliver, 2008). Many researchers have used the Magnusson, Krajcik and Borko (1999) model of PCK for science teaching, and one component of this model is what the authors referred to as orientations to science teaching (OST). In order to support the integrated curriculum and help science teachers improve their teaching performance, there is a need to understand the current PCK status of middle science teachers in Taiwan. Furthermore, we want to know what the most frequently used type of OST is in practical science teaching by most of science teachers. The role of science teaching orientations is associated withhow their PCK components are affected. Our literature review showed that many studies have used the idea of orientations to science teaching in order to focus on the nine categories described by Magnusson et al. (Friedrichsen et al., 2009; Parker & Oliver, 2008; Schwarz & Gwekwerere, 2007; Volkmann, Abell & Zgagacz, 2005). Other studies on the relation of orientations with the other PCK components have presented unclear results (Avraamidou & Zemba-Saul. 2005; Nilsson, 2008). Currently, very few studies have addressed the relationship between orientations and other PCK components.

A considerable number of science education studies have concentrated on the advancement of content-specific PCK. Nevertheless, there is a distinct shortage of science teaching studies examining the association between PCK components and science teaching orientations (Friedrichsen, Driel & Abell, 2011). Schwarz and Gwekwerere (2007) have indicated that "considerable changes in - teachers' pedagogical skills and orientations are often extremely difficult to foster, and success in these areas is critical for reform-oriented science teaching" (p.181). Nilsson (2008) provided examples of how orientations are related to other emerging components of pre-service elementary teachers' PCK. However, most studies are not clear about the relationships between orientations and other PCK components. Friedrichsen et al. (2011) have indicated that limited studies have examined the unclear or absent relationship between orientations and the other model components. Many researchers have used the PCK model of Magnusson et al. as a foundation for science teaching; however, crucial for effective teaching, middle-oriented, teacher-perceived measurements are still lacking. Hence, this research was conducted to make up for such a deficiency. Therefore, in order to survey the current PCK status of middle science teachers and examine the association between PCK components and science teaching orientations, a specific instrument needs to be designed.

Literature Review

PCK Conceptualization and Structure

PCK is critical to student learning and is regarded as the knowledge base of teaching in teacher education (Shulman 1986, 1987). Shulman (1987) described PCK as "the blending of content and pedagogy into an understanding of how particular topics, problems, or issues are organized, represented, and adapted to the diverse interests and abilities of learners, and presented for instruction" (p.8). Carlsen (1999) described PCK as "the particular form of content knowledge that embodies the aspects of content most germane to teachability (p. 137)". There appears to be widespread agreement among education researchers as to the nature of PCK as a unique knowledge domain. PCK refers to specific topics and is, therefore, to be discerned from the general knowledge of pedagogy.

Furthermore, PCK concerns the teaching of specific topics and will, therefore, differ considerably from subject matter knowledge (Van Driel et al., 1998). However, PCK influences the science teacher's knowledge of the subject matter, pedagogy, and the context (Gess-Newsome, 1999).

Several studies have revealed that the particularity of PCK is that its specific knowledge is different from traditional pedagogical and content knowledge (Jang & Chen, 2010; Park, & Oliver, 2008; Van Dijk & Kattmann, 2007). That is, PCK, "the category most likely to distinguish the understanding of the content specialist from that of the pedagogue" (Shulman, 1987, p.8), is highly associated with the knowledge integration and transformation of other types of knowledge such as subject matter knowledge, pedagogical knowledge, and knowledge of context (Jang & Chen, 2010). On the other hand, the further development of PCK highly stresses the importance of knowledge acquisition and its application in practical teaching.

For instance, Park and Oliver (2008) showed that "it is unlikely that teachers acquire PCK first, and then enact it. Rather, knowledge acquisition and knowledge use are interwoven within the context of instructional practices" (p.278). Van Driel, De Jong, and Verloop (2002) stated in their review of the literature on PCK that studies of science teachers' PCK indicate that subject matter knowledge is a prerequisite for the development of PCK and that PCK develops in the actual teaching practice of teachers. That teaching experience is essential for the development of PCK implies that beginning teachers usually have little or no PCK. Jang (2011) further pointed out that a novice teacher's lack of classroom teaching experience inevitably influences their PCK in comparison with that of experienced teachers.

Probably due to the special nature of PCK, there has been no consensus on its definition until now (Hashweh, 2005; Park, & Oliver, 2008; Van Driel et al., 1998). Shulman and others have proposed the domains of teacher knowledge (Carlsen, 1999; Grossman, 1990; Magnusson, Krajcik, & Borko, 1999; Tamir, 1988). The definitions of the components within the different models vary. Grossman (1990) has tried to remedy this situation:

> four general areas of teacher knowledge can be seen as the cornerstones of the emerging work on professional knowledge for teaching: general pedagogical knowledge, subject matter knowledge, PCK, and knowledge of context (p. 5).

Grossman (1990) has further indicated that conceptions of purposes for teaching subject matter, curricular knowledge, knowledge of instructional strategies, and knowledge of students' understanding are the four main ingredients of PCK. Van Driel et al. (1998) also suggested that all scholars agree on Shulman's two knowledge components of PCK: 1) knowledge of representations of subject matter and 2) understanding of specific learning difficulties and student conceptions. Magnusson, Krajcik and Borko (1999) made two changes to Grossman's model. First, due to the key role of evaluation in PCK, one focal point of science-specific PCK is "the addition of a new PCK component—knowledge and beliefs of assessment of scientific literacy" (Friedrichsen et al., 2011, p.361). Moreover, another focal point is to redefine "Grossman's definition of conceptions of purposes for teaching subject matter" (Friedrichsen et al., 2011, p.361) as orientations toward science teaching. Magnusson, Krajcik and Borko (1999) have further suggested that PCK should contain Orientations to Science Teaching, Knowledge of Science Curricula, Knowledge of Students' Understanding of Science, Knowledge of Instructional Strategies, and Knowledge of Assessment of Scientific Literacy.

In order to avoid the confusion that is related to the multiple definitions of PCK terminology, this study adopts the science-specific PCK model of Magnusson et al. (1999).

Orientations to Science Teaching (OST)

According to Magnusson et al. (1999), OST can be described as "knowledge and beliefs about the purposes and goals for teaching science at a particular grade level" (p. 97). Nilsson (2008) added

> Knowledge and beliefs about science teaching and learning guide a teacher's instructional decisions about the organization of activities, the content of student assignments, the use of textbooks and curricular materials, and the evaluation of students' learning (p.1283).

There are considered to be nine types of OST: didactic, academic rigor, process, activity-driven, discovery, conceptual change, project-based science, inquiry, and guided inquiry (Magnusson et al., 1999). First, the goal of didactic orientation is to transmit the facts of science, and the purpose of academic rigor orientation is to represent a particular body of knowledge. Friedrichsen (2002) reviewed the list of science teaching orientations and grouped them into teacher-centered orientations, and those based on reform efforts and associated curriculum projects. Both didactic and academic rigor types are regarded as teacher-centered orientation. Furthermore, because of the influence of educational reform efforts on OST, three types of student-centered orientations that included process, activity-driven, and discovery orientations were proposed in the 1960s (Friedrichsen et al., 2011). The aim of process orientation is to help students develop process skills, for example: observation, classification, measurement, etc. The objective of activity-driven orientation is to facilitate students not only to be familiar with learning content, but also to have "hands-on" experiences. The goal of discovery orientation is to "provide opportunities for students on their own to discover targeted science concepts" (Magnusson et al., 1999, p. 100).

In view of the contemporary reform efforts and curriculum projects, the four additional types of student-centered orientations (conceptual change, project-based science, inquiry, and guided inquiry) have gained increasing attention in recent teacher education studies. The goal of conceptual change orientation is to facilitate the development of scientific knowledge by providing students with explanatory contexts that challenge their naïve conceptions. The aim of project-based science orientation is to involve students in investigating solutions to authentic problems. The goal of inquiry orientation is to represent science as inquiry. The purpose of guided inquiry orientation is to "constitute a community of learners whose members share responsibility for understanding the physical world, particularly aspects with respect to using the tools of science" (Magnusson et al., 1999, p. 100). In order to deal with different teaching situations, it is very likely that a teacher could have multiple orientations in practice (Friedrichsen et al., 2011). Hence, it is important that all nine types of OST should be incorporated into the theoretical framework of this study in order to further verify the association between PCK components and each OST.

Science-Specific PCK Model

In addition to the nine types of OST, Magnusson et al. (1999) have also proposed four other PCK components: knowledge of science curricula (KSC), knowledge of students' understanding of science (KSUS), knowledge of instructional strategies (KIS), and knowledge of assessment of scientific literacy (KASL), as shown in figure 1. First, with respect to science teaching, the KSC indicates the teacher's knowledge of how to manage science teaching materials, and organize relevant science teaching resources in a more comprehensive and logical way. Park and Oliver (2008) indicated that "this component is indicative of teacher understanding of the importance of topics relative to the curriculum as a whole" (p.266). Hanuscin et al. (2011) have pointed out that although KSC is very critical to science education, little attention has been directed to the key role of KSC in practical teaching. They have also noted that science teaching's effectiveness and efficiency could be hampered by improper or insufficient KSC. Moreover, the KSUS, which refers to teachers' knowledge of students' learning styles, prior knowledge and learning capabilities of a subject, is not only regarded as one of the key PCK components, but could also be closely connected with teaching performance and student learning (Stump, 2001).

Third, the KIS has received a lot of attention in previous PCK studies, mainly because it could play a direct role in influencing a teacher's teaching performance, in turn, affecting student learning (Stump, 2001). Magnusson et al. (1999) suggested that the KIS should contain subject-specific strategies and topic-specific strategies. Park and Oliver (2008) added that the subject-specific strategies refer to "general approaches to instruction that are consistent with the goals of science teaching in teachers' minds such as learning cycles, conceptual change strategies, and inquiry-oriented instruction" (p.266). Topic-specific strategies refer to "specific strategies that apply to teaching particular topics within a domain of science" (p.266). Finally, in the science-specific PCK model, Magnusson et al. (1999) have further added KASL to the theoretical framework, mainly because of the critical role of evaluation in science education. That is, whether teachers have sufficient and appropriate knowledge to evaluate student learning could play a key role in the learning outcome of science teaching (Hanuscin et al., 2011).

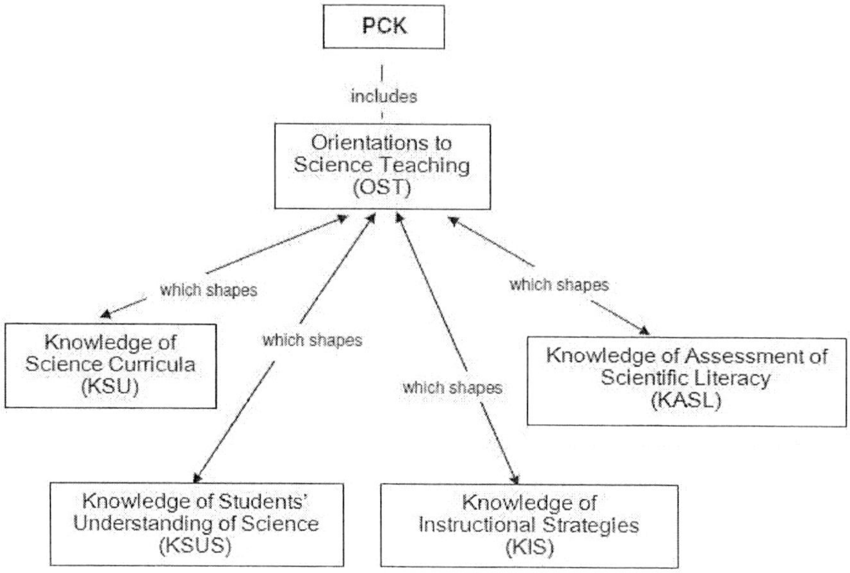

Figure 1. PCK model for science teaching (revised version). Note: From Magnusson, S., Krajcik, J., and Borko, H. (1999). Nature, sources, and development of pedagogical content knowledge for science teaching. In J. Gess-Newsome & N. G. Lederman (Eds.), *Examining Pedagogical Content Knowledge: The Construct and Its Implications for Science Education* (p. 99). Dordrecht, The Netherlands: Kluwer.

More noteworthy, although PCK components are correlated with each other, it has been shown that the advancement of PCK doeas not depend only on the improvement of a single PCK component. Park and Oliver (2008) have already revealed that "if a teacher is unable to integrate the components of the model in a coherent way, improvement within a single component may not be enough to advance PCK and therefore practice" (p.280). Volkmann and Zgagacz (2004) show how science teacher orientations may conflict between a teacher educator and a graduate teaching assistant. However, other PCK components are not investigated. Accordingly, based on previous study suggestions, we use both OST and four PCK components as focal points in this study here.

The Correlation between OST and PCK

Although the model of Magnusson et al. (1999) provides nine orientations and four PCK components, relations between components within the model are either unclear or even absent (Friedrichsen et al., 2011). Avraamidou and Zembal-Saul (2005) used the PCK model of Magnusson et al. to investigate a first-year elementary teacher's specialized practice and knowledge. They conducted an orientation interview to understand the teacher's orientation to teaching and learning science. They focused on the relation between the teacher's knowledge and the practice; however, how this orientation is related to other components of teacher knowledge is not made clear. In the model of Magnusson et al., the orientations component is connected to the other components by two-way arrows as in figure 1, though the connection is merely described as "shaping." The meaning of shaping is not elaborated. That is, if the role of science teachers' OST affects the PCK components or the PCK components affects science teachers' OST, both are not empirically investigated. There is no explanation for what "shaping" means or the relations between the two components. In order to further clarify the link between OST and PCK components, we examined the relationship that exists between all orientations to science teaching and PCK components. We also explore whether the PCK of middle science teachers differs significantly according gender and teaching experience in Taiwan.

METHODOLOGY

Developing the Questionnaire

The questionnaire of the study was designed based on the PCK concept proposed by Magnusson et al. (1999). The PCK concept refers to nine teaching orientations and its relations to the four components (i.e., KSC, KSUS, KIS and KASL). The questionnaire contained three parts. The first part collected information on teachers' gender and teaching experiences and a blank for school teachers to report further comments. In the second part, science teachers needed to select a teaching orientation among nine orientations they used the most in their teaching and then rated the nine orientations on a 5-point Likert scale from 1 (never) to 5 (always). The third part consisted of 33 items on a 5-point Likert scale from 1 (strongly disagree) to 5 (strongly agree), developed for the four PCK components by Magnusson et al. In the study, the items were distributed to A01-A07 in KSC, A08-A14 in KSUS, A15-A25 in KIS, and A26-A33 in KASL as follows.

Magnusson et al. (1999) proposed nine teaching orientations associated with PCK. However, the first three orientations may not be understandable for school teachers. The author elaborated the orientations based on the characteristics listed by Magnusson et al. to better help school teachers understand the first three orientations. The nine teaching orientations in the questionnaire were stated as narrative-based; demonstration and explanation-based; observation, classification and measure-based; activity-driven; discovery; conceptual change; project-based; inquiry; and guided-inquiry. Didactic was elaborated as a narrative-based orientation because teachers who employ this orientation tend to use pedagogical approaches to transmit the scientific facts to students. Academic rigor refers to demonstrations and explanations involved in experiments. Process can refer to developing process skills; therefore, science teachers can employ teaching approaches such as observation, classifications and measures in their science teaching to help students' process scientific concepts through these approaches.

Based on the PCK model of Magnusson et al. (1999), the KSC factor with 7 questions was designed to evaluate teachers' knowledge of curricular materials, curricular saliency, horizontal and vertical curricula. KSC includes science goals and objectives, and specific science curricula. Sample items in this category are:

A1. I know the science curricula I am teaching.
A2. I know the goal of my teaching curriculum.
A3. I know the content of the curriculum well.

A4. I am familiar with the process of how scientific theories or principles have been developed.
A5. I can answer scientific questions students ask.
A6. I know the connection between science curricula and technology.
A7. I can explain the influence of scientific concepts to the society in teaching.

The KSUS factor with 7 questions was developed to measure teachers' knowledge of students' learning requirements, and areas of: student difficulty, including misconceptions, difficulties, prior knowledge, need, interest, and motivation. Sample items in this category are:

A8. I know students' prior knowledge for the science curriculum.
A9. I know students' prerequisites in learning the science curriculum.
A10. I know students' misconceptions related to the science curriculum.
A11. I know students' learning difficulties in the science curriculum.
A12. I know students' learning approaches in learning the science curriculum.
A13. I adjust my teaching methods based on students' learning styles.
A14. I adjust my teaching content based on students' prior knowledge.

The KIS factor with 11 questions was developed to assess teachers' knowledge of subject-specific strategies, and topic-specific strategies, including representations and activities. Sample items in this category are:

A15. I provide opportunities for students to express their views during class.
A16. The pedagogical approaches I use can help students think seriously.
A17. I use different teaching activities to enhance students' learning interests.
A18. I adopt different teaching methods for different course units.
A19. I use familiar examples to explain scientific concepts.
A20. I use appropriate graphs to explain scientific concepts.
A21. I use demonstrations to explain scientific concepts.
A22. I use authentic materials to help students understand the scientific concepts.
A23. I use familiar analogies to explain concepts of subject matter.
A24. I use the language students can understand to explain scientific concepts
A25. I integrate multimedia and computer technology to explain the scientific concepts.

The KASL factor with 8 questions was designed to measure teachers' knowledge of literacy of science learning to assess, and methods of assessing science learning. Sample items in this category are:

A26. I mainly use paper-based tests to evaluate students' understanding of the topic.
A27. My test questions evaluate students' understanding of the topic.
A28. I use different approaches to find out whether students understand the concepts or not.
A29.The assessment methods I design can evaluate students' scientific literacy.
A30. The tests I design can evaluate students' degree in comprehending the subject concepts.

A31. The tests I design can evaluate the degree of students' knowledge in different subject contents.
A32. The tests I design can evaluate students' knowledge of applications.
A33. The tests I design can help students self-evaluate their understanding of the scientific concepts they have learned.

Item Analysis

In order to establish the content and face validity, two professionals in the science education field and ten science teachers reviewed the readability and clarity associated with the narrations of the items. After modifying item descriptions based on the expert feedback, the author used the revised questionnaire for middle school science teachers. The target population of this study was science teachers of middle schools (seventh to ninth grade) in Taiwan. Based on the information from Ministry of Education, this study finally chose a sampling that represented 20 middle schools from northern, central, southern, and eastern Taiwan, and simultaneously gathered the data from each school. There were 225 science teachers responding to the questionnaire. Excluding cases with incomplete data on the items, there were 216 science teachers that remained for establishing the questionnaire's reliability and validity..

The responded questionnaires were grouped with the top 27% and low 27% of the PCK. A T-test was conducted to obtain the critical ratio (CR) and examine the significant differences on each item in the two groups. The critical ratio of item A26, "I mainly use paper-based tests to evaluate students' understanding of the topic", was below the minimum value of .3. All other items indicated significant differences in the two groups at alpha value .05.

The values of the Corrected item-total correlation ranged from .432 to .642 in KSC, from .522 to .666 in KSUS, from .275 to .654 in KIS, and .170 to .600 in the KASL. There were two items (i.e., A25 in KIS and A26 in KASL) that had low values for the corrected item-total which did not satisfy the criterion of the value being over .3. The item description of A25 was "I integrate multimedia and computer technology to explain scientific concepts." The values of the rest items were all over .4, meeting the criteria of being over .3. In addition, the values of Cronbach's alpha if Item Deleted indicated that when these two items (i.e., A25, A26) were deleted, the Cronbach's alpha became higher, demonstrating that A25 and A26 did not measure the same concept that other items measured. Therefore, the items A25 and A26 were deleted. The author used the questionnaire with 31 items for factor analysis.

Factor Analysis

A factor analysis was conducted with 216 middle school science teachers in Taiwan. The factor analysis was set with a fixed number of 4 factors to be extracted as the PCK concept employed in the study consists of four components (Magnusson et al., 1999). At the initial analytical step, 31 items were examined. Several criteria were used to examine the factor analysis for a correlation. Firstly, 30 out of the 31 items correlated at least .3 with at least one other item. The value of the Kaiser-Meyer-Olkin (KMO) of sampling adequacy was .917, meaning that the sample size was adequate for factor analysis. The Bartlett's test was significant, χ^2 (465) = 3667.976, $p < .001$, indicating that the matrix had some relationships between the variables to be included in the analysis. All values of the diagonals of the anti-image correlation matrix were all above .5, satisfying the criteria that its values should be over .5 to include each item in the factor analysis. Additionally, the values of communalities were all above .3, further verifying that each item shared some common variance with other items.

The factor loading of item A15, "I provide opportunities for students to express their ideas during class", was not over .4 so the item was deleted as it did not meet a minimum criterion of having a factor loading over .4. The item A28, "I use different evaluation approaches to evaluate the degree of students' understanding", did not fit the other items extracted for the factor (i.e. KIS). Item A16, "The pedagogical approaches I use can help students think seriously", did not fit the factor extracted with other items for teachers' knowledge on the assessment of scientific literacy. These two items were also deleted as they did not measure what was supposed to be measured, and the deletion did not influence the Cronbach's alpha between the inclusion and exclusion of the items for the two components (i.e., KASL, KIS). The item A24, "I use the language students can understand to explain scientific concept", was originally designed for the knowledge of instructional strategies but was extracted into the component "knowledge of students' science understanding". As this item could be used to measure the two components (i.e., KIS, KSUS), the item was deleted. The remaining items all had factor loadings over .4 in the four components: Knowledge of Science Curricula (KSC, 7 items), Knowledge of Students' Understanding of Science (KSUS, 7 items), Knowledge of Instructional Strategies (KIS, 7 items), and Knowledge of Assessment of Scientific Literacy (KASL, 6 items). In the process, a total of six items were eliminated. The finalized questionnaire is listed in the Appendix.

Second Factor Analysis of the Finalized Questionnaire

There were six items totally eliminated from the original questionnaire designed for the study. A factor analysis was conducted again with the remaining 27 items. A principle components analysis indicated that the four factors totally explained 59.014% of the variance. The initial eigenvalues showed that the first factor explained 37.672% of the variance, the second factor 6.566% of the variance, the third factor 8.577% of the variance, and the fourth factor 6.199% of the variance. The factor loadings of all the items were over .4. The reliability test was constructed and the value of Cronbach's alpha was .875 of KSC, .880 of KSUS, .851 of KIS, and .849 of KASL, indicating high internal consistency. Table 1 provides the values of factor loading for each item of the four components in the finalized questionnaire.

Table 1. Factor Loading of Each Item in the PCK components of the Questionnaire

Item	KSUS	KSC	KIS	KASL
A9	.725			
A11	.710			
A14	.706			
A10	.689			
A8	.682			
A12	.677			
A13	.676			
A3		.855		
A1		.822		
A2		.778		
A5		.594		
A4		.589		
A6		.534		
A7		.453		
A21			.692	
A17			.666	
A18			.652	
A20			.646	
A23			.641	
A22			.629	
A19			.605	
A32				.837
A31				.752
A33				.744
A30				.723
A27				.570
A29				.564

RESULTS

1. Middle School Science Teachers' PCK

Table 2 shows descriptive statistics of the students' responses to the four categories in the questionnaire including mean scores and the standard deviation. The highest mean score regarding PCK is the KSC (M = 4.535), the second highest mean score is the KIS (M = 4.158), the third highest is the KSUS (M = 4.108), and the lowest mean score is the KASL (M = 3.920). As seen in the above results, science teachers considered their Knowledge of Science Curricula rich and positive. However, they indicated that the Knowledge of Assessment of Scientific Literacy could be improved.

Table 2. Mean and Standard Deviation for the Four Components

Components	Mean	S.D.
KSC	4.356	.436
KSUS	4.108	.427
KIS	4.158	.444
KASL	3.920	.446

2. Science Teachers' PCK According to Gender and Teaching Experience

As a number of teachers did not provide information on their gender, teaching experience, or teaching orientations. The author selected 180 teachers to further examine the differences by gender and teaching experiences. Among these teachers, there were 98 male and 82 female science teachers, and their teaching experiences ranged from less than 5 years to more than 26 years.

An independent samples t-test was conducted to examine whether there were gender differences in the teachers' overall PCK and the four components (see Table 3). The results indicated that KSC was found to be significantly different by gender, meaning that male science teachers rated their knowledge of science curriculum significantly higher than female science teachers.

Table 3. Differences of the Four PCK Components by Gender

Components/Group	Male (n = 98)		Female (n = 82)		
	M	S.D.	M	S.D.	*t*
KSC	4.440	.429	4.256	.426	2.880**
KSUS	4.145	.438	4.065	.412	1.253
KIS	4.160	.477	4.155	.403	.078
KASL	3.954	.471	3.880	.413	1.105
PCK	4.176	.357	4.091	.336	1.645

Note. **p < .01.

ANOVA was conducted to examine whether teachers' overall PCK and the four components were significantly different by teaching experiences (see Table 4). Three components (i.e., KSC, KSUS, and KASL) and the overall PCK were found to be statistically significant. Experienced teachers tended to rate their knowledge of science curriculum, knowledge of instructional strategies, and knowledge of assessment of scientific literacy higher than novice teachers.

Table 4. Differences of the Four PCK Components by Teaching Experiences

Components /Group	<5 (n = 31)		6-15 (n = 74)		16-25 (n = 60)		> 26 (n = 15)		
	M	S.D.	M	S.D.	M	S.D.	M	S.D.	*F*
KSC	4.194	.380	4.307	.478	4.474	.398	4.466	.348	3.682*
KSUS	3.926	.441	4.068	.389	4.244	.462	4.143	.254	4.388**
KIS	4.046	.424	4.122	.413	4.226	.486	4.294	.420	1.790
KASL	3.839	.505	3.827	.427	4.005	.393	4.210	.461	4.520**
PCK	4.001	.332	4.091	.351	4.235	.341	4.267	.265	4.575**

Note. *p < .05, **p < .01.

3. Science Teachers' OST and Relation

Table 5 provides information on the percentage of each teaching orientation that the middle school science teachers used the most often in their teaching in Taiwan. Within the same group of participants, science teachers on the top three percentages of teaching orientations, narrative-based (Didactic), demonstration and explanation-based (Academic rigor), and activity-driven. The three orientations were selected to examine its relationships with the four components. A correlation coefficient test was conducted (see Table 6). The results showed that in the narrative-based group, there was a statistically significant correlation between KIS and the orientation; however, its relationship was weak ($r = .191$) and negative, meaning that the higher the teachers rated themselves in applying narrative-based teaching approaches, the lower the teachers' knowledge of instructional strategies was. There were no significant correlations found between the demonstration and explanation-based orientation and the four PCK components. KASL was found to be significantly correlated to the activity-driven orientation. This result indicated that the higher the teachers rated themselves in applying activity-driven teaching approaches, the higher the teachers' knowledge of assessment of scientific literacy was.

Table 5. Teachers' Report on Teaching Orientation They Use the Most Often

	Number of teachers	Percentage (%)
OST		
1. Narrative-based (Didactic)	122	67.8
2. Demonstration and explanation-based (Academic rigor)	20	11.1
3. Process (Observation, classification and measure-based)	7	3.9
4. Activity-driven	14	7.8
5. Discovery	2	1.1
6. Conceptual change	2	1.1
7. Project-based science	4	2.2
8. Inquiry	2	1.1
9. Guided inquiry	7	3.9

Table 6. Pearson Correlations between Orientations and PCK components

Orientation/ components	KSC	KSUS	KIS	KASL
Narrative-based	.075	-.032	-.191*	-.008
Demonstration and explanation	.283	.206	.263	.087
Activity-driven	.377	.233	.305	.540*

Note. *$p < .05$.

DISCUSSION AND IMPLICATIONS

Researchers studying PCK by using the PCK concept of Magnusson et al. (1999) were mainly done by qualitative studies. This study, differing from previous studies, was aimed to develop a questionnaire to gather quantitative data to examine middle school science teachers' teaching orientations and the four PCK components (i.e., KSC, KSUS, KIS, and

KASL) in the context in Taiwan. It was found that the didactic way was the most commonly employed orientation of science teaching in Taiwanese middle schools. This finding was not surprising, because middle school students still had to take the joint entrance exams for placement in high school, and the easiest way for science teachers to cover all the required materials for the entrance exams was to primarily use narratives in teaching. If they would use other student-centered orientations such as discovery, project-based, or inquiry as their main teaching orientation they would not have enough time to cover all the required materials for the entrance exam. The other two teaching orientations science teachers used the most in Taiwan were demonstration and explanation-based (academic rigor), and activity-driven. According to Friedrichsen et al. (2011), the two orientations, didactic and academic rigor, were identified as teacher-centered orientations. Although most teachers in Taiwan employed teacher-centered teaching orientations, there were still few teachers employing student-centered teaching orientations. Therefore, in Taiwan's context, the instructional practices that science teachers used the most in middle schools were narratives, demonstrations and explanations, and hands-on activities.

The results showed that in the narrative-based (didactic) group, there was a statistically significant correlation between KIS and the orientation; however, its relationship was weak and negative, indicating that the higher the teachers rated themselves in applying narrative-based teaching approaches, the lower the teachers' knowledge of instructional strategies was. KIS could be the key component in relation to science teachers' didactic orientation, meaning that when science teachers employed the didactic orientation, they tended to show inadequate knowledge in incorporating instructional strategies into their teaching.

There were no significant correlations found between the demonstration and the explanation-based (Academic rigor) orientation and the four PCK components. KASL was found to be significantly correlated to the activity-driven orientation. When the science teachers used more activities in teaching, they tended to hold more knowledge in using different ways to assess students' science learning as teachers needed to design various forms of assessments for different activities they implemented in classroom. In accordance with the study (Hanuscin et al., 2011), the importance of assessment knowledge is vital for teachers: their effective use and having sufficient knowledge on specific assessment topics can be a channel to provide teachers feedback for continuing to develope their knowledge of their students and pedagogical strategies. Knowledge of assessment plays a key role in evaluating students' understanding since it provides information about students' prior knowledge, misconceptions, mastery level, and learning difficulties. Henze, Van Driel and Verloop (2008) found that science teachers used a variety of ways to assess students' understanding, including: written tests, oral and poster presentations, student reflection, classroom debates, and portfolios. Future research is needed to examine what types of assessments could be effectively used in what kinds of activities and with what specific topics in science teaching and learning.

Teachers' teaching experiences were found to be significantly different in the three components (i.e., KSC, KSUS and KASL) and the overall PCK. Experienced teachers tended to rate their knowledge of science curriculum, knowledge of instructional strategies, and knowledge of assessment of scientific literacy higher than novice teachers. Research on PCK indicates that teachers' development in PCK can be influenced by an individual's teaching experiences (van Driel et al., 2002). Teachers with more teaching experience tend to hold more knowledge on the targeted curriculum and students' understanding of science and know how to assess students' understandings on specific topics. Science teachers must be well-informed about a student's prior knowledge and existing conceptions about the topics to be taught. With such information, science teachers are able to determine a students' readiness to learn, and thus set appropriate goals and materials for their students. Moreover, science teachers must be able to identify a student's area of learning difficulty by being knowledgeable about the topics students find most difficult and the kinds of errors that students commonly make, thus enabling them to modify instructional representations and strategies for their students.

The results also showed that KSC was found to be significantly different by gender, meaning that male science teachers rated their knowledge of science curriculum significantly higher than female science teachers. This could be a focal point in future PCK studies because the KSC could play a key role not only in relating to PCK, but also in influencing student learning (Hanuscin et al., 2011; Park & Oliver, 2008). Science teachers must maintain a broad knowledge base of the targeted curriculum so that they are able to retrieve and teach contents in logical and organized ways. Science teachers' KSC informs designs of horizontal and vertical curricula for a subject, differentiates between core and peripheral concepts, and implement course activities relevant to the course.

In Taiwan's context, it is likely that the implementation of teacher-centered instructional methods could be easier than that of student-centered instructional methods since there is less complexity in PCK knowledge integration and in the processes involved with the teaching and learning environment of the teacher-centered orientation. Since student-centered orientations have received increased attention in recent years, whether science teachers have sufficient PCK is a critical issue in the educational field. It could be challenging for science teachers without sufficient PCK to effectively carry out teaching and learning projects in a student-centered environment, and the unsatisfactory teaching performance could reduce student learning (Hanuscin et al., 2011; Park & Oliver, 2008).

Some researchers have used the model of Magnusson et al. (1999) but ignored the overarching orientation components, focusing on just one or two PCK components (Davis, 2004). Other studies have compared and related two components of the model such as KSUS and KIS, while omitting the orientations component (Brown, Friedrichsen & Abell, 2009; De Jong et al., 2005). However, we could not find the empirical studies to develop an instrument according to the model of Magnusson et al. (1999) and examine the overarching orientations components. Therefore, the developed instrument of this study based on a sound theoretical framework can be used to further examine science teachers' teaching orientations and PCK in various contexts and scientific subjects and with different grades. The study, differing from prior PCK studies with few participants using qualitative methods such as interviews and course portfolios (De Jong, Van Driel & Verloop, 2005; Van Driel, De Jong, & Verloop, 2002; Park & Oliver, 2008), used a quantitative method to examine teachers' PCK and their teaching orientations.

On the whole, the model of Magnusson et al. (1999) includes nine science teaching orientations and four PCK components. From the item analysis, factor analysis and empirical investigation in the Taiwanese middle school science teaching context, the four PCK components (KSC, KSUS, KIS, and KASL) are supported by the theoretical framework (Friedrichsen et al., 2011; Magnusson et al., 1999). Using the questionnaire in different contexts might result in different conclusions. For example, science teachers from elementary schools may be more likely to employ student-centered instructional strategies such as encouraging students to design a project and asking open-ended questions to inquire as to a student's thinking as teachers face less pressure for student's having no need to take entrance exams when entering a middle school. Different results may be obtained by using the developed questionnaire in different countries as the contexts and educational systems can influence how school teachers choose teaching orientations and how their PCK is formed.

However, there were limitations of using the instrument in the study. The results were presented quantitatively in which we could not investigate thorough relationships possibly existent between each teaching orientation and PCK components. For example, science teachers that did not have a suitable and sufficient KIS component could have a tendency to prefer thr didactic orientation. In addition to using the questionnaire, the researchers may select several case teachers to collect qualitative data in various ways, such as interviews, observations, and portfolios to reveal their tendency of science teaching orientations and PCK.

Conclusion

Although the science-specific model of Magnusson et al. (1999) often appears in current science education studies, previous studies have not yet designed an instrument to evaluate current science teachers' teaching orientation and their PCK in middle schools in Taiwan. The results of this study have made contributions to the current research in science education in Taiwan by examining current science teachers' teaching orientation and the correlations between PCK components and several teaching orientations. Though the student-centered orientation has gradually become a mainstream teaching and learning method in the current educational system, we found that Taiwanese middle school science teachers still prefer using teacher-centered orientations. Therefore, practitioners and professionals in the science education field in Taiwan can find the findings useful when they further reveal evidence in helping science teachers develop their teaching from teacher-centered toward student-centered teaching orientations.

References

Abell, S.K. (2008). PCK twenty years later: Does it remain a useful idea? *International Journal of Science Education, 30*, 1405–1416.

Avraamidou, L. & Zembal-Saul, C. (2005). Giving priority to evidence in science teaching: A first-year elementary teacher's specialized practices and knowledge. *Journal of Research in Science Teaching, 42*, 965 –986.

Brown, P., Friedrichsen, P., &Abell, S. (2009, April). Teachers' knowledge of learners and instructional sequencing in an alternative certification program. Paper presented at the annual meeting of the American Educational Research Association, San Diego, CA.

Carlsen, W. S. (1999). Domains of teacher knowledge. In J. Gess-Newsome, & N. G. Lederman (Eds.), Examining pedagogical content knowledge (pp. 133–144). Dordrecht: Kluwer.

Davis, E. (2004). Knowledge integration in science teaching: Analyzing teachers' knowledge development. *Research in Science Education, 34*, 21 – 53.

De Jong, O., Van Driel, J.H. & Verloop, N. (2005). Preservice teachers' pedagogical content knowledge of using particle models in teaching chemistry. *Journal of Research in Science Teaching, 42*(8), 947–964.

Friedrichsen, P. (2002). A substantive-level theory of highly-regarded secondary biology teachers' science teaching orientations. Unpublished doctoral dissertation, The Pennsylvania State University, University Park.

Friedrichsen, P., Abell, S., Pareja, E., Brown, P., Lankford, D., & Volkmann, M. (2009). Does teaching experience matter? Examining biology teachers' prior knowledge for teaching in an alternative certification program. *Journal of Research in Science Teaching, 46*, 357 – 383.

Friedrichsen, P., Van Driel, J. H. & Abell, S. K. (2011). Taking a closer look at science teaching orientations. *Science Education, 95*(2), 358-376.

Gess-Newsome, J. (1999). Pedagogical content knowledge: An introduction and orientation. In J. Gess-Newsome, & N. G. Lederman (Eds.), Examining pedagogical content knowledge (pp. 3–17). Dordrecht: Kluwer.

Gess-Newsome, J. & Lederman, N.G. (1993). Preservice biological teachers' knowledge structures as a function of professional teacher education: A year-long assessment. *Science Teacher Education, 77*(1), 25-43.

Grossman, P. L. (1990). The making of a teacher: Teacher knowledge and teacher education. New York: Teachers College Press.

Hanuscin, D. L., Lee, M. H. & Akerson, V. L. (2011). Elementary teachers' pedagogical content knowledge for teaching the nature of science. *Science Education, 95*(1), 145-167.

Hashweh, M. Z. (2005). Teacher Pedagogical Constructions: A reconfiguration of pedagogical content knowledge. *Teachers and Teaching: Theory and Practice, 11*, 273–292.

Henze, I., Van Driel, J. & Verloop, N. (2008). Development of experienced science teachers' pedagogical content knowledge of models of the solar system and the universe. *International Journal of Science Education, 30*(10), 1321-1342.

Jang, S. J. (2006). The effects of incorporating web assisted learning with team teaching in seventh-grade science classes. *International Journal of Science Education. 28*(6), 615-632.

Jang, S. J. (2009). PCK and Teaching Innovations. NY: Nova Science Publishers, Inc.

Jang, S. J. (2011). Assessing college students' perceptions of a case teacher's pedagogical content knowledge using a newly developed instrument. *Higher Education, 61*(6), 663-678.

Jang, S. J. & Chen, K. C. (2010). From PCK to TPACK: Developing a transformative model for pre-service science teachers. *Journal of Science Education and Technology, 19*(6), 553-564.

Loughran, J.J., Mulhall, P. & Berry, A. (2004). In search of pedagogical content knowledge in science: Developing ways of articulating and documenting professional practice. *Journal of Research in Science Teaching, 41*, 370–391.

Magnusson, S., Krajcik, J. & Borko, H. (1999). Nature, sources and development of pedagogical content knowledge for science teaching. In J. Gess-Newsome & N. G.

Lederman (Eds.), *Examining pedagogical content knowledge* (pp. 95 – 132). Dordrecht, The Netherlands: Kluwer.

Nilsson, P. (2008). Teaching for understanding: The complex nature of pedagogical content knowledge in preservice education. *International Journal of Science Education, 30*(10), 1281 – 1299.

Park, S. & Oliver, J. S. (2008). Revisiting the conceptualization of pedagogical content knowledge (PCK): PCK as a conceptual tool to understand teachers as professionals. *Research in Science Education, 38*, 261–284.

Schwarz, C. V. & Gwekwerere, Y. N. (2007). Using a guided inquiry and modeling instructional framework (EIMA) to support preservice K-8 science teaching. *Science Education, 91*(1), 158-186.

Shulman, L. S. (1986). Those who understand: knowledge growth in teaching. *Educational Researcher, 15*, 4-14.

Shulman, L. S. (1987). Knowledge and teaching: foundations of the new reform. *Harvard Educational Review, 57*, 1-22.

Stump, S. L. (2001). Developing pre-service teachers' pedagogical content knowledge of slope. *The Journal of Mathematical Behavior, 20*(2), 207-227.

Tamir, P. (1988). Subject matter and related pedagogical knowledge in teacher education. *Teaching and Teacher Education, 4*, 99–110.

Van Dijk, E. M. & Kattmann, U. (2007). A research model for the study of science teachers' PCK and improving teacher education. *Teaching and Teacher Education, 23*(6), 885-897.

Van Driel, J.H., Verloop, N. & de Vos, W. (1998). Developing science teachers' pedagogical content knowledge. *Journal of Research in Science Teaching, 35* (6), 673 – 695.

Van Driel, J.H., De Jong, O. &Verloop, N. (2002). The development of pre-service chemistry teachers' PCK. *Science Education, 86*, 572–590.

Volkmann, M., Abell, S. & Zgagacz, M. (2005). The challenges of teaching physics to preservice elementary teachers: Orientations of the professor, teaching assistant, and students. *Science Education, 89*, 847 – 869.

Volkmann, M. & Zgagacz, M. (2004). Learning to teach physics through inquiry: The lived experiences of a graduate teaching assistant. *Journal of Research in Science Teaching, 41*, 584 – 602.

Appendix A: Science Teachers' PCK Assessment Questionnaire

Finalized PCK Questionnaire

Draw a circle around

1. if teaching practice takes place Never (N.)
2. if teaching practice takes place Seldom (Se.)
3. if teaching practice takes place Sometimes (So.)
4. if teaching practice takes place Often (O.)
5. if teaching practice takes place Always (A.)

Orientations to Science Teaching (OST)		N	Se	So	O	A
1	I mainly use narrative-based way in my science teaching.	①	②	③	④	⑤
2	I mainly use demonstration and explanation-based way in my science teaching.	①	②	③	④	⑤
3	I invite my students to observe, classify and measure in science teaching.	①	②	③	④	⑤
4	I mainly use an activity-driven teaching in my science class.	①	②	③	④	⑤
5	I mainly use discovery learning in my science class.	①	②	③	④	⑤
6	I mainly use conceptual change way or curriculum in my science teaching,	①	②	③	④	⑤
7	I mainly use project-based science learning in my class.	①	②	③	④	⑤
8	I mainly use inquiry learning in my science class.	①	②	③	④	⑤
9	I mainly use guided-inquiry approach in my science class.	①	②	③	④	⑤
10	From the above question 1-9, which way do you most frequently use OST type in practical science teaching.	1 6	2 7	3 8	4 9	5

Knowledge of Science Curricula (KSC)	
A01	1. I know the scientific contents I teach.
A02	2. I know the goals of the contents I teach.
A03	3. I am familiar with the subject contents on the topics I teach.
A04	4. I am familiar with the process of how scientific theories or principles have been developed.
A05	5. I can answer scientific questions students ask.
A06	6. I know the connection between science curricula and technology.
A07	7. I can explain the influence of scientific concepts to the society in teaching.
Students' Understanding of Science (KSUS)	
A08	1. I know students' prior knowledge for the science curriculum.
A09	2. I know students' prerequisites in learning the science curriculum.
A10	3. I know students' misconceptions related to the science curriculum.
A11	4. I know students' learning difficulties in the science curriculum.
A12	5. I know students' learning approaches in learning the science curriculum.
A13	6. I adjust my teaching methods based on students' learning styles.
A14	7. I adjust the teaching content based on students' prior knowledge.
Knowledge of Instructional Strategies (KIS)	
A17	1. I use different teaching activities to enhance students' interests in learning.
A18	2. I use different teaching approaches to teach different course units.
A19	3. I use familiar examples to explain scientific concepts.
A20	4. I use appropriate graphs to explain scientific concepts.
A21	5. I use demonstrations to explain scientific concepts.
A22	6. I use authentic materials to help students understand scientific concepts.
A23	7. I use analogies that students are familiar with to explain scientific concepts.
Knowledge of Assessment of Scientific Literacy (KASL)	
A27	1. The questions I ask can evaluate students' understanding of the topic.
A29	2. The assessment methods I design can evaluate students' scientific literacy.
A30	3. The tests I design can evaluate students' degree in comprehending the subject concepts.
A31	4. I tests I design can evaluate the degree of students' knowledge in different subject contents.
A32	5. The tests I design can evaluate students' knowledge of applications.
A33	6. The tests I design can help students self-evaluate their understanding of the scientific concepts they have learned.

Chapter 2

DEVELOPING AN INSTRUMENT FOR EVALUATING UNIVERSITY STUDENTS' PERCEPTIONS OF TEACHERS' PCK

ABSTRACT

The ongoing professional development for university teachers has been much emphasized. However, previous research has seldom addressed university students' perceptions associated with the knowledge of their teachers. The purpose of this study was to develop an instrument that could be employed to evaluate university students' perceptions of teachers' Pedagogical Content knowledge (PCK). According to the pilot test results, the four categories finally constructed were Subject Matter Knowledge (SMK), Instructional Representation and Strategies (IRS), Instructional Objects and Context (IOC), and Knowledge of Students' Understanding (KSU). In the main study, seven items were generated under each category, making a total of 28 items in the questionnaire, and the instrument was administered to 172 college education students. Analysis results indicated that the instrument about teachers' knowledge show satisfactory validity and reliability. Suggestions for the instrument's application in future research are also made.

Keywords: PCK; University teachers' knowledge; Professional development; University students' perceptions; Evaluation instrument

INTRODUCTION

The notion of ongoing professional learning and development for college teachers has been much emphasized (Clarke & Hollingsworth, 2002; Fullan & Stiegelbauer, 1991; Garcia & Roblin, 2008; Guskey, 1985; Johnson, 1993). The central focus of current professional development efforts aligns most closely with the changes associated with a growth in the learning perspective. Within this perspective, change is identified with learning, and it is regarded as a natural and expected component of the professional activity of teachers and schools (Clarke & Hollingsworth, 1994). Many novice college teachers with doctoral degrees and a certain level of their subject matter knowledge were not able to be effective teachers

(Clarke & Hollingsworth, 2002; Jang, 2008; Major & Palmer, 2006). One of the reasons was that they did not need to get a Teacher's Certificate to become primary and secondary school teachers (Jang, 2008). Aside from the subject matter knowledge, they needed more pedagogical knowledge (Hasweh, 1987; Leinhardt & Smith, 1985; Lenze & Dinham, 1994). There are many ways for a college teacher to gain professional growth, such as: researching, instructional observance and reflection, publication of reports and books, joining workshops to share experiences with others, seminars, attending professional activities, journal reading, writing, curriculum designing, and peer activities (Ball, 1996; Cooney & Krainer, 1996; Dalgarno & Colgan, 2007; Lieberman, 1995; Loucks-Horsely et al., 2003; Sykes, 1996). Nevertheless, promoting college teachers' "Pedagogical Content Knowledge (PCK)" is the key point to advancing the professional growth of teachers (Lenze & Dinham, 1994).

It has also been reported that the success of college teaching depends not only on the teacher's subject-matter knowledge, but also on their personal understanding of their students' prior knowledge and learning difficulty (Grossman, 1990; Lederman, Gess-Newsome & Latz, 1994). In addition, other factors of success include their own teaching methods and strategies, curriculum knowledge, educational situation, goal and value (Shulman, 1987). In particular, the college teachers' pedagogical content knowledge is the main issue of the current college education revolution. Shulman's notion of PCK has attracted much attention and has been interpreted in different ways (Geddis, Onslow, Beynon & Oesch, 1993; Grossman, 1990). The foundation of science PCK is thought to be the amalgam of a teacher's pedagogy and understanding of content such that it influences their teaching in ways that will best engender a student's learning for understanding. Initially, college teachers separate subject-matter knowledge from general pedagogical knowledge. These types of knowledge are, however, being integrated as a result of teaching experiences. By getting acquainted with the specific conceptions and ways, teachers may start to restructure their subject-matter knowledge into a form that enables productive communication with their students (Lederman, Gess-Newsome and Latz, 1994). According to Lederman, Gess-Newsome and Latz (1994), the development of PCK among science teachers is promoted by the constant use of subject-matter knowledge in different teaching situations. Many scholars suggest that PCK is developed through an integrative process rooted in classroom practice, and that PCK guides a teacher's actions when dealing with a specific subject matter in the classroom.

Purpose of the Research

Greater emphasis has been paid on the development and research of elementary and secondary teachers' PCK (Dalgarno & Colgan, 2007; De Jong, Van Driel & Verloop, 2005; Gess-Newsome & Lederman, 1993; Grossman, 1990; Loughran, Mulhall & Berry, 2004; Van Driel et al., 1998). Some studies have even developed an instrument for examining students' perceptions of a secondary teacher's knowledge (Tuan, Chang, Wang & Treagust, 2000). However, previous research on learning environments has seldom addressed college students' perceptions of teachers. The purpose of this research was to develop an instrument for evaluating college students' perceptions of teachers' PCK in order to help college teachers better understand how they teach.

THEORETICAL FRAMEWORK

Pedagogical Content Knowledge

The impact of constructivist epistemology seems to be important in PCK. As constructivism emphasizes the role of previous experience in knowledge construction processes, it is not surprising that a teacher's knowledge is studied in relation to their practice in research from this point of view. Shulman (1987) regarded PCK as the knowledge base for teaching. This knowledge base comprises seven categories, three of which are content-related (content knowledge, PCK, and curriculum knowledge). The other four categories refer to general pedagogy, learners and their characteristics, educational contexts, and educational purposes. PCK is concerned with the representation of concepts, pedagogical techniques, knowledge of what makes concepts difficult or easy to learn, knowledge of students' prior knowledge, and theories of epistemology. It also involves knowledge of teaching strategies that incorporate appropriate conceptual representations in order to address a learner's difficulties and misconceptions while fostering meaningful understanding (Mishra & Koehler, 2006).

Reynolds (1992) summarizes the literature on PCK in the following aspects: (1) teaching students through different skills and methods; (2) thinking about the content scope and sequence that needs to be covered; (3) understanding students' previous conceptions, skills, knowledge and interests related to the particular topic; (4) using appropriate representation to introduce subject matter knowledge; (5) using different strategies to help students understand and become interested in a topic; (6) using appropriate evaluation methods to assess students' understanding of subject matter knowledge. Tuan (1996) further investigated the essence of PCK and suggested that the components of PCK include a teacher's understanding of the subject matter knowledge, their methods, teaching representations, curriculum knowledge, assessment knowledge, knowledge of students' understanding of the topics, and knowledge of the context of the learning environment (Tuan, Chang, Lee, Wang & Cheng, 2000).

Grossman (1990) has tried to remedy this situation by distinguishing four general areas of teacher knowledge that can be seen as the cornerstones of the emerging work on professional knowledge for teaching: general pedagogical knowledge, knowledge of context, subject matter knowledge, and PCK. Within Grossman's knowledge base for teaching, the general pedagogical knowledge is defined as knowledge concerning general principles of instructions, learning and learners, knowledge related to classroom management, and knowledge about the aims and purposes of education. Knowledge of context includes knowledge of the school setting, for example culture, and knowledge of individual students (Van Dijk & Kattmann, 2007). PCK is a unique domain that is informed by other knowledge areas. There seems to be a reciprocal relationship between PCK and the foundational knowledge domains, subject matter, pedagogy, and context. The foundational knowledge domains inform PCK, influencing the teacher's knowledge of the subject matter, pedagogy, and the context (Gess-Newsome, 1999).

So far, several PCK-related conceptualizations have been presented; however, we do not intend to be exhaustive. Instead we intend to bring to light to the fact that "there is no universally accepted conceptualization of PCK" (Van Driel, Verloop, & de Vos, 1998, p. 677) due to its amorphous nature. Furthermore, as the world is constantly changing and impacting education, components that constitute teacher knowledge are subject to change. Nevertheless, Van Driel et al. (1998) suggested that all scholars agree on Shulman's two knowledge components: 1) *knowledge of representations of subject matter* and 2) *understanding of specific learning difficulties and student conceptions*. Here, we also found that Shulman's other two knowledge components are more often cited: 3) *subject matter knowledge* and 4) *knowledge of educational contexts*. It seems that the above four components *(knowledge of representations of subject matter, student learning difficulties, subject matter knowledge,* and *knowledge of educational contexts)* are more fundamental, constituting the core of pedagogical content knowledge in this study.

Major and Palmer (2006) used a qualitative study of faculty members participating in a university campus-wide, problem-based learning initiative to examine the process of transforming faculty pedagogical content knowledge. They found that the faculty's existing knowledge and institutional intervention influenced new knowledge of faculty roles, student roles, disciplinary structures, and pedagogy. A teacher's PCK is deeply personal, highly contextualized and influenced by teaching interactions and experiences (De Jong, Van Driel & Verloop, 2005; Van Dijk & Kattmann, 2007; Van Driel, Beijaard & Verloop, 2001). Mulholland and Wallace (2005) suggested that science a teacher's pedagogical content knowledge requires the longitudinal development of experience as they develop from novices into experienced teachers. Certain studies also showed that a teacher well equipped with the subject-matter knowledge might be able to transfer his/her knowledge in an efficient way, making it easier for students to acquire knowledge (Carter & Doyle, 1987; Tobin & Garnett, 1988). The subject-matter structures of biology teachers were investigated during a year of professional teacher education (Gess-Newsome & Lederman, 1993). Their knowledge structures appeared to be mainly derived from the college science courses. While these structures were often vague and fragmented on entering teacher education, they developed toward more coherent and integrated views of biology during teacher education.

Students' Perceptions of Teachers' Knowledge

Jang (2009) designed a peer coaching-based model for in-service science teachers' PCK. Four science teachers and 123 secondary students took part in this study on the application of the developed PCK-RIER model (Research, Instruction, Evaluation, and Reflection). Students thought that science teachers had a rich subject matter knowledge and often assessed knowledge of their understanding. However, science teachers had difficulty implementing a representational repertoire and instructional strategies. Students' perceptions and teachers' reflection are the important factors of this model for developing a science teacher's PCK. It is recommended that this model should be adopted in teacher education to offer more opportunities for professional growth among science teachers.

Knight and Waxman (1991) pointed out that although a student's perceptions might not be consistent with the reality generated by outside observers, they could present the range of reality for individual students and their peers in the classroom. Using a student's perceptions could enable researchers and teachers to appreciate the perceived instructional and environmental influences on their learning processes. According to Lloyd and Lloyd (1986), students expected teachers to provide a sense of how the constituent parts of a discipline fit together, to have rich and adequate subject matter knowledge, and to be able to teach this subject matter knowledge to their understanding level.

Olson and Moore (1984) revealed that, from the student's perspective, a good teacher knows the subject matter knowledge well, explains things clearly, makes the subject interesting, gives regular feedback, and gives extra help to students. Similarly, Turley (1994) found that a student's perception of effective teaching was a combination of method, context, student effort, and teacher commitment. Students considered those teachers who knew their subject, showed evidence of thoughtful planning, used appropriate teaching strategies, instructional and representational repertoires, and gave adequate structure and direction to effective teachers (Tuan, Chang, Wang & Treagust, 2000). In brief, research on a student's perception associated with their teacher's teaching revealed that students expect teachers to have strong content knowledge, inferring that they are able to perceive whether teachers' content knowledge was good or bad. Students also expected teachers to use effective instructional methods; in other words, they expected teachers to have good pedagogical content knowledge (Shulman, 1986, 1987).

RESEARCH METHODOLOGY

In order to explore college teachers' understanding of PCK and actualities, we first designed a questionnaire on a novice teacher's PCK. The categories of the questionnaire were constructed from Shulman's (1987) PCK. It comprised 15 questions involved in three categories (instructional representation, strategies, and assessments of students' prior knowledge), and each category had five questions. We conducted a pilot test on 16 novice teachers of PCK workshops at the beginning of the first semester of 2007 in order to examine their understanding of PCK. Moreover, we also selected 182 college students to join the test. After interviewing with some teachers and reviewing the suggestions from the Advancing Teachers' Teaching Excellence Committee (ATTEC), we found many overlaps between instructional representation and strategies. On the other hand, most college teachers put emphasis on subject matter knowledge, and hence instructional context was neglected in this study.

Therefore, we constructed four components from the literature review and pilot study to constitute the core of pedagogical content knowledge in this study. The four categories finally constructed were Subject Matter Knowledge (SMK), Instructional Representation and Strategies (IRS), Instructional Objects and Context (IOC), and Knowledge of Students' Understanding (KSU) (See Appendix). Seven items were generated under each of the four categories as agreed by the researchers and team teachers; the instrument was also revised according to the suggestions from five experienced college teachers of the ATTEC. Once the conceptual framework for the instrument was established, the items should be easy for college students in Taiwan to comprehend. Each category of items should be meaningful from the students' perspective.

Subject Matter Knowledge (SMK) refers to a student's perception of the extent to which the teacher demonstrates a comprehension of the subject matter and ideas within the discipline. The construction process of content knowledge and entire structure and direction of subject knowledge are also included. Examples of items in this category are:

a1. My teacher knows the content he/she is teaching.
a7. My teacher knows the whole structure and direction of this subject matter knowledge.

Instructional Representation and Strategies (IRS) refers to a student's perception of the extent which the teacher uses a representational repertoire including analogies, metaphors, examples, and explanations, and the teacher selects teaching strategies if they benefit the content learning, including informational technology. Examples of items in this category are:

b1. My teacher uses familiar examples to explain concepts related to the subject matter.
b7. My teacher uses multimedia or technology (e.g. PowerPoint) to express the concept of subject.

Instructional Objects and Context (IOC) comprises knowledge about the aims and process of education. IOC also includes the interactive atmosphere in the curriculum, teachers' attitudes, knowledge related to classroom management, knowledge of the school setting, and instructional values.

c1. My teacher makes me clearly understand the objectives of this course.
c6. My teacher appropriately copes with our classroom context.
c7. My teacher's teaching belief or value is active and aggressive.

Knowledge of Students' Understanding (KSU) refers to the college student's perception of the extent to which the teacher evaluates student understanding before and during interactive teaching, and at the end of lessons and units. Examples of items in this category are:

d1. My teacher realizes students' prior knowledge before the class.
d7. My teacher's tests help me realize the learning situation.

Analysis of Data

The pilot study indicated that the survey yielded high reliability and validity results (Jang, Guan, & Hsieh, 2009), and some items from the original instrument were revised. The revised instrument representing college students' opinions about their teacher's PCK were divided into four main categories, namely SMK, IRS, IOC and KSU, consisting of seven items per category for a total of 28 items. The questionnaire survey employed in this research included one open-ended question for students to comment on the course. Finally, participants of this study were students in the College of Humanities and Education at the University. They took courses from the novice college teachers of participating in PCK workshops at the beginning of this semester. A total of 172 valid responses were collected.

This research used the quantitative research method. Statistical analyses on the survey data were carried out. The survey adopted the Likert scales, with five points designed for students to express their opinions as follows: "Never", "Seldom", "Sometimes", "Often", and "Always" correspond respectively to 1 - 5 points according to students' responses. Moreover, we also tested reliabilities and validities in connection with the questionnaire. For reliabilities,

Cronbach's alpha would be adopted to evaluate the internal consistency. On the other hand, a factor analysis would be adopted to evaluate the construct validities.

RESULTS AND DISCUSSION

Table 1 shows the descriptive statistics of the students' responses to the four categories in the questionnaire including mean scores and the standard deviation. Results obtained by ANOVA show significant differences ($p = .000 < .05$, $F = 46.757$) among the four categories. The highest mean score is the SMK ($M = 4.35$, $SD = 0.542$), followed by the IRS ($M = 4.28$, $SD =0.573$) and the IOC ($M = 4.23$, $SD = 0.634$) with the score of the KSU ($M = 3.99$, $SD = 0.828$) being the lowest. All of them indicate that these items were investigating occurrences between "often" and "always". As seen in the above results, students considered their teachers' subject matter knowledge (SMK) rich and positive, but indicated that knowledge of students' understanding (KSU) could be improved.

Table 1. Descriptive Statistics of the College Students' Responses for Four Categories

Category	Items	N	Mean	SD
SMK	7	172	4.35	0.542
IRS	7	172	4.28	0.573
IOC	7	172	4.23	0.634
KSU	7	172	4.00	0.664

Furthermore, we also analyzed each item. Items with a mean below 4 points included: Q-b3 (My teacher's teaching methods keep me interested in this subject), Q-d1 (My teacher realizes students' prior knowledge before the class.), Q-d2 (My teacher knows student's learning difficulties of this subject before the class.), and Q-d4 (My teacher's tests evaluate my understanding of the subject.). The study showed that college students expressed that most of the teachers could not fully understand students' difficulties and prior knowledge (KSU). The critical way to promote KSU was to make students become aware of their teachers' understanding and pay attention to their learning difficulties. From students' comments, it may be caused by teachers' subjectivities, a faster progress of the program, and too many assignments, affecting teachers' teaching results. Having too many students in a normal class was also a problem. We advised to adopt an e-learning method to make up for the insufficient interaction among students and college teachers.

Reliabilities of the Instrument

In that reliability of this instrument, we used Cronbach alpha values to evaluate its internal consistency. After statistical analyzing, the Cronbach's alpha value of 28 items was 0.965, indicating that the questionnaire had a good internal consistency. Moreover, in "Item-

Total Statistics" (Table 2), the section of "Corrected Item-Total Correlation" revealed each corrected item not only presented a higher correlation (all correlation values were greater than 0.400), but also had a higher psychological homogeneity. Furthermore, from the section of "Cronbach's alpha if item Deleted", it meant: if we deleted one of the items, the Cronbach's alpha values were almost the same, even smaller than the deleted items. It followed from what had been said that the instrument had a higher consistency and reliability so that it was unnecessary to delete any item from the scale.

Table 2. Item-Total Statistics

Item	Scale Mean if Item Deleted	Scale Variance if Item Deleted	Corrected Item-Total Correlation	Cronbach's Alpha if Item Deleted
a1 (SMK01)	113.39	226.508	.543	.965
a2 (SMK02)	113.51	223.000	.674	.964
a3 (SMK03)	113.58	222.982	.673	.964
a4 (SMK04)	113.81	217.875	.772	.963
a5 (SMK05)	113.72	220.205	.736	.964
a6 (SMK06)	113.83	218.983	.649	.964
a7 (SMK07)	113.56	221.253	.721	.964
b1 (IRS01)	113.64	221.518	.649	.964
b2 (IRS02)	113.83	218.948	.726	.964
b3 (IRS03)	114.06	214.184	.745	.964
b4 (IRS04)	113.76	218.197	.647	.964
b5 (IRS05)	113.59	222.208	.607	.964
b6 (IRS06)	113.80	218.795	.767	.963
b7 (IRS07)	113.26	227.747	.484	.965
c1 (IOC01)	113.69	219.641	.737	.964
c2 (IOC02)	113.70	218.865	.770	.963
c3 (IOC03)	113.74	216.756	.803	.963
c4 (IOC04)	113.90	214.492	.826	.963
c5 (IOC05)	113.83	217.502	.663	.964
c6 (IOC06)	113.88	216.669	.797	.963
c7 (IOC07)	113.56	222.750	.652	.964
d1 (KSU01)	113.99	218.403	.674	.964
d2 (KSU02)	114.15	217.353	.726	.964
d3 (KSU03)	113.94	218.242	.725	.964
d4 (KSU04)	113.99	217.357	.741	.963
d5 (KSU05)	113.90	219.891	.651	.964
d6 (KSU06)	113.96	217.852	.667	.964
d7 (KSU07)	113.97	218.543	.639	.964

*	Item a1-a7 means question 1 – 7 of SMK, Item b1-b7 means question 1 – 7 of IRS, Item c1-c7 means question 1 – 7 of IOC, Item d1-d7 means question 1 – 7 of KSU.

After exploring the item-total statistics, the factor analysis of 28 items would be the next in order to test the construct validity. By factor analyzing, we tried to examine whether constructs of this questionnaire were consistent with the four perspectives of PCK that we defined or not. Besides, we also checked the definitions and concepts of those items which were involved in four categories to find if they were accurate or not. Finally, the questionnaire was formed so that we could use it to do the research and explore further.

Validity

The validity of the new instrument was confirmed in terms of its content and construct validity. Content validity refers to the extent that the content of the items measure what is claimed to be measured (Anastasi, 1988; Zeller, 1994); in this case, content validity was ascertained by responses during the process of developing the instrument from college students and experienced college teachers. Additionally, the researchers in teachers' pedagogical content knowledge areas provided comment and critical feedback. According to Zeller (1994), construct validity focuses on the assessment of whether a particular measure relates to the measures consistent with a theoretically anticipated way and this is usually done by factor analysis (Anastasi, 1988). Based on the factor loadings, the items in each category were deleted that had low item-scale correlations from the reliability section and that had factor loadings less than 0.40.

Wu (2007) mentioned that we could separately adopt a factor analysis with categories if: 1. The pilot questionnaire had cited some theories and references associated; 2. Items were clearly divided into a few categories (perspectives); 3. All of the items were also clearly defined; 4. All of the items were checked and revised by some experts. On the other hand, we could prevent some problems such as too many extracted components, having extracted components and contents that were irrelevant to the original design, or too many items needing to be deleted.

The Factor Analysis of Four Categories: SMK, IRS, IOC, and KSU

In table 3, the factor loadings of seven items in the SMK were over 0.600 (ranging from 0.762 to 0.860), and the percentage of its total variance explained was 64.514%. It revealed that the construct could explain the variance of items up to 36% at least, and all of the items in the SMK were retained. For the IRS, the factor loadings of its seven items were over 0.600 (ranging from 0.625 to 0.819), and the percentage of its total variance explained was 57.031%. It also revealed that the construct could explain the variance of items up to 36% at least, and all of the items in the IRS were retained. For the IOC, the factor loadings of its seven items were over 0.600 (ranging from 0.745 to 0.885), and the percentage of its total variance explained was 67.659%. It also revealed that the construct could explain the variance of items up to 36% at least, and all of items in the IOC were retained. Finally, the factor loadings of seven items in the KSU were over 0.600 (ranging from 0.749 to 0.834), and the percentage of its total variance explained was 64.159%. It also revealed that the construct could explain the variance of items up to 36% at least, and all of the items in the KSU were retained. After a factor analysis with the 4 categories, the items of each category were approximately confirmed. If the numbers of distributed items is appropriate, we had better proceed to further testing the internal consistency of four categories. After re-examining internal consistency of the items in the four categories, we found the Cronbach's alpha value of the four categories still to be higher (ranging from .871 to .918). In other words, items of each category had a higher internal consistency.

Table 3. Factor analysis of the 28-item instrument in four categories based on data from 172 college students in Taiwan

Items No.	Factor Loading Subject Matter Knowledge	Instructional Representation & Strategies	Instructional Objective & Context	Knowledge of Students' Understanding
a3 (SMK03)	.860			
a2 (SMK02)	.847			
a7 (SMK07)	.823			
a5 (SMK05)	.809			
a4 (SMK04)	.808			
a6 (SMK06)	.739			
a1 (SMK07)	.726			
b2 (IRS02)		.819		
b1 (IRS01)		.785		
b4 (IRS04)		.777		
b6 (IRS06)		.775		
b3 (IRS03)		.772		
b5 (IRS05)		.719		
b7 (IRS07)		.625		
c3 (IOC03)			.885	
c4 (IOC04)			.881	
c2 (IOC02)			.860	
c6 (IOC06)			.857	
c1 (IOC01)			.769	
c7 (IOC07)			.747	
c5 (IOC05)			.745	
d3 (KSU03)				.834
d2 (KSU02)				.818
d1 (KSU01)				.804
d6 (KSU06)				.804
d7 (KSU07)				.803
d4 (KSU04)				.792
d5 (KSU05)				.749

Conclusion

The major contribution of this study was to develop an instrument that could be used to evaluate a college student's perceptions of their teachers' PCK. The findings differ from Major and Palmer (2006) when they used qualitative methods, such as interviews and course portfolios, to describe the process of existing and transforming the faculty's PCK. In our study, we adopted the quantitative way (survey) to analyze and realize the PCK that college faculty members (especially novice teachers) would have. The data analysis indicated that the questionnaire of this study had satisfactory validity and reliability. The uniqueness of the survey was that it was specifically related to a college teacher's knowledge within particular teaching and learning context. This was important because research had shown a college teacher's knowledge influenced a student's perceptions of the learning environment. Tobin (1996) showed that teachers' metaphors and beliefs influenced how they taught and implemented the curriculum, and their level of content knowledge influenced whether or not students were taught for factual retention or for understanding. Tobin and Fraser (1990) also investigated the teaching characteristics of exemplary science teachers. These effective teachers not only used management strategies to sustain students' engagement, but also teaching strategies such as problem solving activities. Furthermore, they also provided concrete examples for abstract concepts, asked questions to increase students' understanding, helped students to engage in both large or small group activities, and maintained favorable classroom learning environments.

In this study, college students expressed that most teachers were insensible to a student's learning difficulties and their prior knowledge, which became the weak point of the college teacher's teaching. Concerning the teaching strategy, college teachers adopted an inefficient teaching approach, which couldn't stimulate students' interests of learning for a certain subject matter. The causes which formed students' comments were: teachers' subjective attitude; their explanatory lecturing in class; and the speedy procession of their curriculum, which couldn't match up with students' learning status. The current questionnaire functioned as an instrument to help college teachers actively understand their students' reaction toward the procession of courses, improve the comprehension on students' learning difficulties, further adopt proper and efficient teaching strategies and achieve a better performance in teaching. Through the reflective research and participation to the PCK workshops, college teachers would amend their teaching approaches and strategies, and further improve their ability in course design (Dalgarno & Colgan, 2007; Loucks-Horsely et al., 2003; Sykes, 1996).

The new perspective in further research concerned the implementation of the questionnaire as an instrument to investigate a college teacher's teaching results based on PCK. Participants were college teachers from different departments on the campus, not only from the College of Humanities and Education (as done previously). Upon the analysis of the survey collected from university students' feedback in class as well as the follow-up testing of reliability and validity of the collected questionnaire, the college teacher's auto-examination mechanism was established on the teaching performance and the efficiency related to PCK. An investigation of the questionnaire was planned to be implemented in the middle and the end of semester. Other data sources were collected from personal interviews with teachers, as well as from students' feedback information, and further examined individual teacher's improvement of PCK. Furthermore, it could also be designed as a teaching or research model for college science teachers to develop their PCK (Jang, 2011). The above-mentioned helped to comprehend the interactive condition between college teachers and students in class, constituting a substantial reference for college curriculum design and instruction in the future. (The article has been revised from WCES 2009).

REFERENCES

Anastasi, A. (1988). *Psychological testing* (6th ed.) New York: Macmillan.

Ball, D. L. (1996). Teacher learning and the mathematics reforms: What we think we know and what we need to learn. *Phi Delta Kappan, 77*, 500–508.

Carter, K. & Doyle, W. (1987). Teachers' knowledge structure and comprehension process. In J. Calder-head(Ed.), *Exploring teachers' thinking* (pp.147-160). London; Cassel.

Clarke, D. J. & Hollingsworth, H. (1994). Reconceptualising teacher change. In G. Bell, B. Wright, N. Leeson, & J. Geake (Eds.), Challenges in mathematics education: Constraints on construction, Vol. 1. Proceedings of the 17th annual conference of the Mathematics Education Research Group of Australasia (pp. 153–164). Lismore, NSW: Southern Cross University.

Clarke, D. J. & Hollingsworth, H. (2002). Elaborating a model of teacher professional growth. *Teaching and Teacher Education, 18*, 947-967.

Cooney, T. J. and Krainer, K. (1996) Inservice mathematics teacher education: The importance of listening. *International Handbook of Mathematics Education*, (pp. 1155 – 1185). Netherlands: Kluwer Academic Publishers.

Dalgarno, N. & Colgan, L. (2007). Supporting novice elementary mathematics teachers' induction in professional communities and providing innovative forms of pedagogical content knowledge development through information and communication technology. *Teaching and Teacher Education, 23*, 1051-1065.

De Jong, O. Van Driel, J.H., & Verloop, N. (2005). Preservice teachers' pedagogical content knowledge of using particle models in teaching chemistry. *Journal of Research in Science Teaching, 42*(8), 947–964.

Fullan, M. & Stiegelbauer, S. (1991). The new meaning of educational change (2nd ed.). New York: Teachers College Press.

Garcia, L. M. & Roblin, N. P. (2008). Innovation, research and professional development in higher education: Learning from our own experience. *Teaching and Teacher Education, 24*, 104–116.

Geddis, A. N., Onslow, B., Beynon, C. & Oesch, J. (1993). Transforming content knowledge: Learning to teach about isotopes. *Science Education, 77*(6), 575-591.

Gess-Newsome, J. (1999). Pedagogical Content Knowledge: An introduction and orientation. In J. Gess-Newsome & N. G. Nederman (Eds.). *Examining pedagogical content knowledge* (pp.3-17). Dordrecht, The Netherlands: Kluwer.

Gess-Newsome, J. & Lederman, N.G. (1993). Preservice biological teachers' knowledge structures as a function of professional teacher education: A year-long assessment. *Science Teacher Education, 77*(1), 25-43.

Grossman, P.L. (1990). The making of a teacher: Teacher knowledge and teacher education. New York: Teachers College Press.

Guskey, T. R. (1985). Staff development and teacher change. *Educational Leadership, 42*(7), 57-60.

Hasweh, M. (1987). Effects of subject matter knowledge in the teaching of biology and physics. *Teaching and Teacher Education, 3*(1), 109-120.

Jang, S. J. (2008). The effects of integrating technology, observation and writing into a teacher education method course. *Computers & Education, 50*(3), 853-865.

Jang, S. J. (2009). Developing in-service science teachers' PCK through a peer coaching-based model. *Journal of Education Research, 3*(1), 87-108.

Jang, S. J., Guan, S. Y. & Hsieh, H.F. (2009). Developing an instrument for assessing college students' perceptions of teachers' pedagogical content knowledge. *Procedia Social and Behavioral Sciences, 1* (1), 596–606.

Jang, S. J. (2011). Assessing college students' perceptions of a case teacher's pedagogical content knowledge using a newly developed instrument. *Higher Education, 61*(6), 663-678.

Johnson, N. (1993). A celebration of teachers as learners. Paper presented at The Australian College of Education 1993 National Conference: Global Economy, Global Curriculum. Melbourne.

Knight, S. L. and Waxman, H. C. (1991). Analyzing effective teaching of Hispanic students' problem-solving strategies in Spanish. *NABE Annual Conference Journal, 1988-1989*. Washington, D.C.: National Association for Bilingual Education.

Lederman, N.G., Gess-Newsome, J. & Latz, M.S. (1994). The nature and development of preservice science teachers' conceptions of subject matter and pedagogy. *Journal of Research in Science Teaching, 31*, 129–146.

Leinhardt, G. & Smith, D. (1985). Expertise in mathematics instruction: Subject matter knowledge. *Journal of Educational Psychology, 77*(3), 247–271.

Lenze, L.F. & Dinham, S. M. (1994). Examining pedagogical content knowledge of college faculty new to teaching. Paper presented at the Annual Meeting of the American Educational Research Association, New Orleans, Louisiana.

Lieberman, A. (1995). Practices that support teacher development. *Phi Delta Kappan, 76*, 591–596.

Lloyd, B. C. and Lloyd, R. C. (1986) Teaching/learning: The student viewpoint. *Reading Horizons, 26*, 266-269.

Loucks-Horsely, S., Love, N., Stiles, K., Mundry, S. & Hewson, P. W. (2003). Designing Professional Development for Teacher of Science and Mathematics, 2nd edition. Thousand Oaks, CA: Corwin Press, Inc.

Loughran, J.J., Mulhall, P. & Berry, A. (2004). In search of pedagogical content knowledge in science: Developing ways of articulating and documenting professional practice. *Journal of Research in Science Teaching, 41*, 370–391.

Major, C. & Palmer, B. (2006). Reshaping teaching and learning: The transformation of faculty pedagogical content knowledge. *Higher Education, 51*, 619–647.

Mishra, P. & Koehler, M. J. (2006). Technological pedagogical content knowledge: A new framework for teacher knowledge. *Teachers College Record. 108*(6), 1017-1054.

Mulholland, J. & Wallace, J. (2005). Growing the tree of teacher knowledge: Ten years of learning to teach elementary science. *Journal of Research in Science Teaching, 42*(7), 767 – 790.

Olson, L. & Moore, M. (1984). Voices form the classroom: Students and teachers speaking out on the quality of teaching in our schools. (ERIC Document Reproduction Service No. ED 252 497.)

Reynolds, A. (1992). What is competent beginning teaching? A review of the literature. *Review of Educational Research, 62*(1), 1-35.

Shulman, L.S. (1986). Those who understand: Knowledge growth in teaching. *Educational Researcher, 15*, 4–14.

Shulman, L.S. (1987). Knowledge and teaching: Foundations of the new reform. *Harvard Educational Review, 57*, 1–22.

Sykes, G. (1996). Reform of and as professional development. *Phi Delta Kappan, 77,* 465-4676.

Tobin, K. (1996). Teacher learning, teacher education and educational reform. In Treagust, D.F., Duit, R., and Fraser, B.J. *Improving Teaching and learning in science and mathematics* (pp. 175-189). New York: Teachers College Press.

Tobin, K. & Fraser, B. J. (1990). What does it mean to be an exemplary science teacher? *Journal of Research in Science Teaching, 27*(1), 3-25.

Tobin, K. & Garnett, P.(1988). Exemplary practice in science classroom. *Science Education, 72*, 197-208.

Tuan, H. L. (1996). Investigating the nature and development of pre-service chemistry teachers' content knowledge, pedagogical knowledge and pedagogical content knowledge. *Proceedings of National Science Council, Part D, Math-Science-Technology Education, 6*(2), 101-112.

Tuan, H.L., Chang, W.H., Lee, C.K., Wang, C.Y. & Cheng, P.Y. (2000). Developing a Pedagogical Content Competence Evaluation for apprentice physical science teachers – A case study. Proceedings of the NSC-Part D: *Mathematics, Science, and Technology Education, 10*(1), 1-14.

Tuan, H-L., Chang, H-P., Wang, K-H. & Treagust, D. F. (2000). The development of an instrument for assessing students' perceptions of teachers' knowledge. *International Journal of Science Education, 22*(4), 385-398.

Turley, S. (1994), *The way teachers teach is, like, totally whacked: The student voice on classroom practice.*, American Educational Research Association (AERA) in New Orleans, U.S.A.

Van Dijk, E. M. & Kattmann, U. (2007). A research model for the study of science teachers' PCK and improving teacher education. *Teaching and Teacher Education, 23*, 885-897.

Van Driel, J. H., Beijaard, D. & Verloop, N. (2001). Professional development and reform in science education: The role of teachers' practical knowledge. *Journal of Research in Science Teaching, 38*(2), 137 – 158.

Van Driel, J. H., Verloop, N. & De Vos, W. (1998). Developing science teachers' pedagogical content knowledge. *Journal of Research in Science Teaching, 35*, 673–695.

Wu, M. L. (2007). SPSS operation and application: The practice of quantitative analysis of questionnaire data. Taipei, Taiwan: Wu Nan.

Zeller, M., (1994). Determinants of credit rationing: A study of informal lenders and formal credit groups in Madagascar. *World Development, 22*(12), 1895-1907.

APPENDIX: ASSESSING STUDENTS' PERCEPTIONS OF UNIVERSITY TEACHERS' PCK

Directions for students:

This questionnaire contains five statements about teaching practices which could take place in this class. You will be asked how often each practice takes place. There are no "right" or "wrong" answers. Your opinion is what is wanted. Think about how well each statement describes what this class is like for you. You will be asked to freely describe your personal comments to the course in the end. Be sure to give an answer for all questions.

Draw a circle around

1. if teaching practice takes place Never
2. if teaching practice takes place Seldom
3. if teaching practice takes place Sometimes
4. if teaching practice takes place Often
5. if teaching practice takes place Always

A. SMK (Subject Matter Knowledge)		C. IOC (Instructional Objective & Context)	
1	My teacher knows the content he/she is teaching.	1	My teacher makes me clearly understand objectives of this course.
2	My teacher explains clearly the content of the subject.	2	My teacher provides an appropriate interaction or good atmosphere.
3	My teacher knows how theories or principles of the subject have been developed.	3	My teacher pays attention to students' reaction during class and adjusts his/her teaching attitude.
4	My teacher selects the appropriate content for students.	4	My teacher creates a classroom circumstance to promote my interest for learning.
5	My teacher knows the answers to questions that we ask about the subject.	5	My teacher prepares some additional teaching materials.
6	My teacher explains the impact of subject matter on society.	6	My teacher copes with our classroom context appropriately.
7	My teacher knows the whole structure and direction of this SMK.	7	My teacher's belief or value in teaching is active and aggressive.
B. IRS (Instructional Representation & Strategies)		D. KSU (Knowledge of Students' Understanding)	
1	My teacher uses appropriate examples to explain concepts related to subject matter.	1	My teacher realizes students' prior knowledge before class.
2	My teacher uses familiar analogies to explain concepts of subject matter.	2	My teacher knows students' learning difficulties of subject before class.
3	My teacher's teaching methods keep me interested in this subject.	3	My teacher's questions evaluate my understanding of a topic.
4	My teacher provides opportunities for me to express my views during class.	4	My teacher's assessment methods evaluate my understanding of the subject.
5	My teacher uses demonstrations to help explaining the main concept.	5	My teacher uses different approaches (questions, discussion, etc.) to find out whether I understand.
6	My teacher uses a variety of teaching approaches to transform subject matter into comprehensible knowledge.	6	My teacher's assignments facilitate my understanding of the subject.
7	My teacher uses multimedia or technology (e.g. PowerPoint) to express the concept of subject.	7	My teacher's tests help me realize the learning situation.

Comments:

In this course, if you have any learning difficulty or opinion, please describe it as follows:

____

____

____

Chapter 3

DEVELOPING UNIVERSITY TEACHERS' PCK THROUGH A REFLECTIVE TEACHING MODEL AND STUDENTS' PERCEPTIONS ON TEACHING

ABSTRACT

Previous research on learning environments has seldom addressed a university teacher's PCK. This research aimed to evaluate how a reflective model could affect a university student's perceptions of their teacher's PCK development. A developed survey and reflective model were the tools for six novice teachers to facilitate reflection and exchange ideas among teachers. The survey was conducted twice in this study; with the pre-test carried out during the mid-term examination, and the post-test administered in the last week of the semester. This study adopted a case study approach and conducted both quantitative and qualitative analyses. The results show that except for Teachers B and C, who showed no significant difference in scores between the two surveys, the other four teachers significantly improved. We also compared, in detail, the students' perception of six novice teachers' PCK performance, individually in four dimensions. In particular, this study clearly describes each teacher's teaching performance before and after the reflection. The research limitations and difficulties encountered are also discussed in this study.

Keywords: College teaching and learning; PCK; Reflective model; Higher education; Professional development

INTRODUCTION

Considering a novice university instructor with doctoral degrees and possessing a certain level of their subject content knowledge, the instructor must also know how to translate that content knowledge into meaningful instruction for his students. This specialized knowledge has been labeled pedagogical content knowledge (PCK) and, when integrated into a college teacher's subject matter, this then promotes a teacher's teaching efficiency (Shulman, 1986, 1987). Most university instructors entering the profession may find their initial teaching efforts stressful, but they acquire a repertoire of teaching strategies and representations, with experience, that they draw on throughout their teaching. While an instructor's style of teaching provides a means of coping with many of the routine demands of teaching, there is also a danger that it can hinder his/her professional growth. The university teachers' PCK is the key to advancing the professional growth of teachers (Lenze & Dinham, 1994) and is an important indicator for assessing the professional skills of university teachers (Jang, 2009a). In fact, PCK has been described as a knowledge base necessary for effective teaching in many educational reform documents (American Association for the Advancement of Science [AAAS, 1993]; National Research Council [NRC, 1996].

Shulman (1987) regarded PCK as the knowledge base for teaching. This knowledge base comprises seven categories, three of which are content related (subject-matter knowledge, PCK, and curriculum knowledge). The other four categories refer to general pedagogy, learners and their characteristics, educational contexts and educational purposes. PCK combines subject-matter content and pedagogy, suggesting that this amalgam represents the understanding necessary for transforming subject matter into forms or viable instructions that are more accessible to students (Abell, 2008). It can be effectively and flexibly used in the communication process between teachers and learners during classroom practice. The crucial factor in this development of PCK is teaching reflection and experience (De Jong, Van Driel & Verloop, 2005; Loughran, Mulhall & Berry, 2004; Tuan, Chang, Wang & Treagust, 2000). The central goal of reflection is to develop a teacher's understanding of why they employ certain instructional strategies and how they can improve their teaching to have a positive effect on students (Lee, 2005). A reflection process can be a development activity as well as a means of enhancing the PCK of instructors. Reflective practice is becoming the favored paradigm for continuing professional development in higher education (Clegg, Tan & Saeidi, 2002).

In general, a survey to test the teaching effectiveness of university teachers is usually conducted at the end of each semester with the aim to find out the teaching performance of teachers for the previous period of time, and for them to make appropriate modifications accordingly. However, as most of the subjects in the current university curriculum are only taught for one semester, teachers do not have time to make modifications if students' feedbacks are only made available at the end of the semester. In this way, they can only improve their teaching in the next semester. On the other hand, subjects taught in the next semester are usually new ones with new contents. Hence, improvement in teaching the same subject has to be postponed again. When improvement in teaching has been postponed for a lengthy period of time, it may result in poor teaching effectiveness. To overcome the limitations on curriculum design and provide teachers with opportunities to make timely modifications, this questionnaire should be surveyed in the middle of the semester. Such method can not only satisfy formative evaluation requirements, but also have the effects of a diagnostic evaluation with the aim to help university teachers in improving their teaching effectiveness.

Greater emphasis has been put on the research and development of elementary and secondary teachers' PCK (Dalgarno & Colgan, 2007; De Jong, Van Driel & Verloop, 2005; Gess-Newsome & Lederman, 1993; Jang, 2009b; Loughran, Mulhall & Berry, 2004; Van Driel et al., 1998). However, previous research on learning environments has seldom addressed the university teacher's PCK. This research aimed to evaluate how a reflective model could affect a university student's perceptions of their teachers' PCK development. This study employed the "Questionnaire on university students' perception of teachers'

PCK", developed by Jang, Guan and Hsieh (2009), as the research tool. In order to improve teaching and promote reflection, the questionnaire was employed at both mid-term and final examinations. In this model, a PCK workshop and teaching observation were also held for university teachers to facilitate the exchange of ideas among teachers.

THEORETICAL FRAMEWORK

It has also been reported that the success of college teaching depends not only on the teachers' subject-matter knowledge, but also on their personal understanding of students' prior knowledge and learning difficulty (Grossman, 1990; Lederman, Gess-Newsome & Latz, 1994). In addition, other factors of success include their own teaching methods and strategies (Fernandez-Balboa & Stiehl, 1995; Hasweh, 2005; Lenze & Dinham, 1994). The pedagogical knowledge about certain topics and teaching strategies, including the knowledge of representation (as model and metaphor) and activities (as experiment and explanation), were closely related and demand a flexible schema for implementation (De Jong, Van Driel & Verloop, 2005; Grossman, 1990; Lederman, Gess-Newsome & Latz, 1994; Van Dijk & Kattmann, 2007). More importantly, when dealing with pedagogical knowledge, teachers' actions will be determined, to a large extent, by their PCK, making PCK an essential component of professional knowledge. Some studies also showed that a science teacher well equipped with the subject-matter knowledge might be able to transfer his/her knowledge in a more efficient way, enabling the students to receive the knowledge more easily (Carter & Doyle, 1987; Tobin & Garnett, 1988). When teaching unfamiliar topics, science teachers express more misconceptions (Hashweh, 1987) and they talk longer and more often, posing questions of a low cognitive level (Carlsen, 1993). These results are interpreted in terms of PCK rather than subject-matter knowledge (Sanders, Borko & Lockard, 1993).

Magnusson et al. (1999) stressed that the development of PCK is determined by the content to be taught, the context in which the content is taught, and the way the teacher reflects on his/her teaching experiences. Although teaching experience was taken as the primary source of PCK development, reflection was also a critical component of professional development. Reflection therefore emerges as another important element for novice teachers in developing expertise in their practice, and is central to their accepting more responsibility for their actions (Loughran, 2002; Shulman, 1987; Wallace & Oliver, 2003). Nilsson (2008) emphasized the role of teaching experience and reflection as a way of better understanding the complex entities that constitute a knowledge base for teaching. She draws attention to the value of teachers participating in experiences that might contribute to the development of their PCK and supports the view of PCK development as a process of transformation. Mezirow (1990) stated that reflection is an examination of the justification for one's beliefs, primarily to guide action and to reassess the efficacy of the strategies and procedures used in the classroom. Reflective practice involves the process of teaching and the thinking behind it, rather than simply evaluating the teaching itself.

Major and Palmer (2006) used a qualitative study of faculty members participating in a university campus-wide, problem-based learning initiative to examine the process of transforming faculty PCK. They found that the existing knowledge of faculty and institutional intervention influenced the new knowledge of faculty roles, student roles, disciplinary structures, and pedagogy. Teachers' PCK is deeply personal, highly contextualized and influenced by teaching interaction and experience (Van Driel, Beijaard & Verloop, 2001; De Jong, Van Driel & Verloop, 2005; Van Dijk & Kattmann, 2007). Mulholland and Wallace (2005) suggested that teachers' pedagogical content knowledge requires the longitudinal development of experience as they develop from novices into experienced teachers. Hammersley-Fletcher and Orsmond (2005) used peer observation of teaching as the process employed within higher education establishments that can be instrumental in developing the reflective practices of professional lecturers. Reflection leads to self-knowledge, and this is important to the professional development of instructors in higher education institutions.

Rhine and Bryant (2007) argued to develop teachers' reflective practice by blending online discussion and the use of video of teaching two approaches. In the meantime, they also generalized five themes as a framework for identifying teacher reflection. Finally, the study showed that the use of digital video in web-based discussion was not only an effective way to remain connected to facilitate professional dialogue and collaboration among their students, but also provided a means for them to learn how to reflect-in-action. Clarke (2009) tried to use video recordings and posted their recordings within online discussion areas to help student teachers reflect their teachings. According to questionnaire results , student teachers thought there were some pros and cons. Among the pluses, they found that video-recording could save time and that video clips were more "honest", conveying emotions and complexity in a way that is not possible. On the other hand, it also allowed for more intimacy, liveliness and entertainment.

Penny and Coe (2004) pointed out that reflection has become a common feature in universities across the world, employing student feedback as an indicator of teaching quality. Although students' perceptions might not be consistent with the reality generated by outside observers, they could present the range of reality for individual students and their peer in the classroom (Knight & Waxman, 1991). Students' comments and perspectives can allow university teachers and researchers to bring changes to the teaching, the teaching environment and the students' learning experience as well as provide university teachers with appropriate support to enhance their teaching effectiveness.

The main advantage of applying the "Questionnaire on university students' perception of teachers' PCK" is that it addresses how teachers tend to be trapped in the self-righteous thinking, as they are embodiments of authority in classroom (Jang, Guan & Hsieh, 2009). To observe the teaching scenario more objectively, it is necessary to discuss from the students' viewpoints to better understand the PCK development of teachers. In particular, as the students are involved in the teaching process, whether teachers have achieved the expected pedagogical objectives can be better understood through students' perception. Moreover, it is being judged by all students instead of a small number of observers (De Jong, Van Driel, & Verloop, 2005; Major & Palmer, 2006; Van Driel, De Jong, & Verloop, 2002).

RESEARCH METHOD

This study adopted a case study approach and conducted both quantitative and qualitative analyses. The quantitative data were collected via a survey administered to three classes in a university. As for qualitative data, the researchers analyzed the students' feedbacks on the open-ended questions, teachers' reflection through seminars and individual teacher interviews.

Context and Participants

The research was conducted from September 2008 to January 2009. The participants were several novice teachers at a medium-sized university in Taoyuan, Taiwan. The researchers recruited novice teachers from various colleges who were interested in and were willing to improve their PCK teaching effectiveness. There were six novice university teachers who enrolled in the PCK workshops. The purpose of organizing PCK workshops for the teachers was to gather novice teachers and professionals for discussion and exchange ideas. In this study, a set of workshops were offered to novice teachers with diverse contents of themes. The subjects were students from six classes of the College of Human & Education, the College of Computer Science and the College of Engineering of a university enrolled in the first semester of 2008, and three new teachers (denoted by A-F respectively) (see Table 1). Table 1 shows the number of students taking the course, and the number of valid samples collected before and after the survey.

Table 1. Description of the Classes, Teachers and Participants with Different Subject Majors

College	Teacher	Subject	Total	Valid samples
HE	A	Chinese	62	47
	B	Introduction to Chinese culture	64	48
	C	Information systems & Internet foundations	62	28
EECS	D	Introduction to computer science	65	46
	E	Probability & statistics	60	50
E	F	Engineering mathematics	71	41

Research Design and Procedure

In this study, students surveys on their perceptions of university teachers' PCK, teaching recordings, teacher interviews, and PCK workshops were mainly used to improve the teachers' reflection on teaching. The reflective model of college teaching (see Figure 1) was designed by the researchers and comprised several steps as follows: (1) Initial instruction; (2) Teaching recording and observation; (3) Midterm assessment; (4) Final assessment; (5) Interviewing & teaching revising, and the center part; and (6) Teaching Reflection (which continuously proceeded during the above stages).

As seen in Figure 1, teaching reflection (step 6) is the main focus of this model and keeps proceeding from step 2 to step 5. Step 1 is the initial instruction. Through the first workshop, researchers illustrate and discuss with the teachers who participate in this study. In addition, researchers introduce the related theories of PCK and the research tools in this study. In step 2, researchers discuss the time and date of the teaching recording with the case teachers in advance. Then the assistants go to the class and record about one to two courses. After recording in the class, the video is transferred to DVD. This can provide the case teachers with an opportunity to observe and reflect on their teaching. Step 3 is the midterm assessment. Researchers use the questionnaires to survey students' perceptions of the teacher's PCK and then analyze the results of the questionnaires. During step 3 to step 4, the second workshop is arranged in order to realize the students' perceptions through the results of pre-test for teacher's PCK. Then researchers discuss how to promote teachers' teaching with the case teachers. Step 4 is the final assessment. Researchers apply the questionnaires again as the post-test to investigate the difference of the teacher's PCK after the teaching reflection. During step 4 to step 5, the third workshop is held in order to illustrate the analysis of the questionnaires among the pre-test and post-test and discuss students' perceptions of the teacher's PCK. In step 5, researchers interview with the case teachers and discuss the results based on the video, pre-test, and post-test. Finally, researchers would ask for the teacher's teaching reflection after watching the DVD (step 6), and then brainstorm with each other to provide the case teachers with some suggestions to achieve better teaching.

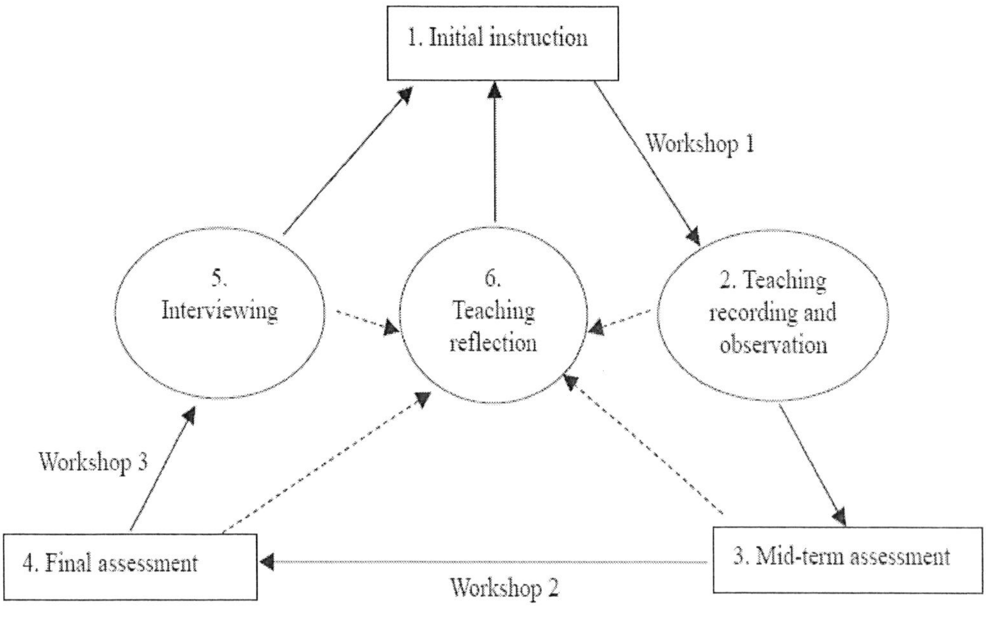

Figure 1. The reflective model of university teaching.

Data Collection and Tools

This study used the "Questionnaire on university students' perception of teachers' PCK" designed and developed by Jang, Guan and Hsieh (2009) as the tool (see Appendix). The contents include four dimensions, which are: Subject Matter Knowledge (SMK), Instructional Representation & Strategies (IRS), Instructional Objects & Context (IOC)) and Knowledge of Students' Understanding (KSU) (see Appendix). There are seven sub-topics for each dimension; thus, there are a total of 28 items. The pilot study analyzed the questionnaire results of 182 college students with a total of 172 valid samples collected. With regard to validity and reliability, the questionnaire presented very high validity and reliability. The reliability analysis was analyzed by internal consistency reliability. The analysis results showed that the Cronbach's α value is 0.965, indicating that the internal consistency of the 28 items is good. Regarding the validity analysis, it was conducted by expert review on PCK-related fields, and factor analysis for verification and modification. In this study, the first and second surveys were conducted during the middle and at the end of the semester, respectively, to further understand the changes in students' perception of teachers' PCK.

The goal of PCK workshops is mainly to gather new university teachers from various colleges and educational research experts for discussion and exchanging views to enhance teaching improvements for better effects through peer group stimulation and conversation. Teachers with excellent teaching performance abilities and educational expertise are invited during the seminar period to provide teaching consultation to new teachers who lack teaching experience. Through interaction and discussion among teachers, they can continuously reflect on the changes in various dimensions of their PCK. Thus, teaching deviations can be modified quickly and effectively once they learn about their students' needs. Before the midterm exam, two research assistants go to the classes for the video recording. One observes the teacher's teaching and takes some notes. The other does the video recording of the teacher's teaching in the classroom. Then, the video is transferred to DVD as a teaching reflection. The individual teacher interviews are mainly presented by teaching diagnosis reports and results of other associated qualitative data. The first-hand feedback information of students is given to teachers to understand the compliance of students' perception with the teaching and so they can immediately conduct reflective teaching.

Data Analysis

Data was collected, summarized and coded by survey, interview and PCK workshops. The quantitative data includes the results of two surveys. The survey data was measured by a Likert 5-point scale, and rated from 1 to 5 points according to the degree of difference. Finally, a comparative analysis of the two surveys was conducted by a t-test. The qualitative data includes the feedbacks of teachers, as well as the continuous comparative discussion on the other qualitative data. The qualitative analysis is carried out according to Patton (1990): 1) gather all the original data; 2) organize, categorize and edit the original data into files that can be easily identified and acquired; 3) summarize and identify important indexes for in-depth analysis according to study problems and types. In other words, the aim of the data analysis is to allow for a continuous interpretation and explaination of this data (Erickson, 1986). A triangulation method is employed to interpret the histories and changes in the PCK professional growth of the individual teachers. Then, the data is categorized and coded according to different interview contents to establish specific data login categories.

Research Results

Table 2 shows the changes in the average means and standard deviations of six new university teachers in the two surveys. Except for Teachers B and C, who show no significant difference between the two surveys and slightly lower average points, the average points of the other four teachers significantly improve. However, Teacher B's average points in the two surveys are the highest ($M>4.40$), whereas those of Teacher C are relatively low.

Table 2. Performance of Six Novice College Teachers between Pre- and Post- Tests

Teacher	*N*	*M*	*SD*	*t*
A				
Pre-test	47	3.89	1.01	-3.458**
Post-test	47	4.25	0.40	
B				
Pre-test	48	4.55	0.36	1.580
Post-test	48	4.48	0.37	
C				
Pre-test	28	3.75	0.25	0.292
Post-test	28	3.72	0.76	
D				
Pre-test	46	3.87	0.51	-14.662***
Post-test	46	4.23	0.53	
E				
Pre-test	50	3.87	0.54	-5.782***
Post-test	50	3.98	0.60	
F				
Pre-test	41	3.23	0.39	-6.103***
Post-test	41	3.56	0.71	

Notes. ** $p < 0.01$; *** $p < 0.001$.

For a further discussion of the reasons associated with the 6 novice teachers' PCK change, we make a teaching diagnosis and offer suggestions. We also compare, in detail, the students' perception of these 6 novice teachers' PCK performance individually, in four dimensions. The paired sample t-test was conducted to analyze the statistics, comparing whether there were differences between the teachers' scores obtained in the pre-test and post-test.

1.SMK (Subject Matter Knowledge)

Table 3 shows 6 novice teachers' change in the SMK dimension and that 5 of them achieved a level of significant difference from pre-test to post-test, except teacher C. The mean scores of teacher A, D, E and F, which were low in the pre-test, also obviously enhanced in the post-test.

Table 3. Description of 6 novice college teachers' scores on SMK between the pre- and posttest

Teacher	N	Pre-test		Post-test		t
		Mean	SD	Mean	SD	
A	47	3.97	1.03	4.35	0.41	-3.306**
B	48	4.64	0.40	4.48	0.37	2.753**
C	28	3.92	0.45	3.78	0.80	1.232
D	46	4.06	0.78	4.27	0.79	-3.284**
E	51	3.96	0.54	4.15	0.59	-2.875**
F	41	3.53	0.52	3.79	0.11	-3.061**

2. IRS (Instructional Representation & Strategies)

Table 4 shows 6 novice teachers' change in the IRS dimension and that 3 (teachers A, D and F) of them achieved a level of significant difference from pre-test to post-test. The mean scores of teachers C and E, which were low in the pre-test, also enhanced in the post-test, but they did not achieve a level of significant difference, except where teacher F's mean score is much lower than other teachers in the post-test.

3. IOC (Instructional Objective & Context)

Table 5 shows 6 novice teachers' change in the IOC dimension and that 4 (teachers A, B, D and F) of them achieves a level of significant difference from pre-test to post-test. The mean scores of teachers A, D and F in the IOC dimension, which were low in the pre-test, were greatly enhanced in the post-test, except where teacher B's post-test mean score was lower than the pre-test. Teacher E's mean score in the post-test was higher than the pre-test, but did not achieve a level of significance. Teacher C's post-test mean score was lower than in the pre-test.

Table 4. Description of 6 novice college teachers' scores on IRS between the pre- and posttest

Teacher	N	Pre-test		Post-test		t
		Mean	SD	Mean	SD	
A	47	3.93	0.95	4.26	0.43	-3.359**
B	48	4.48	0.40	4.43	0.41	0.746
C	28	3.60	0.38	3.69	0.73	-0.787
D	46	3.66	0.57	4.18	0.56	-8.713***
E	51	3.84	0.61	3.94	0.66	-1.646
F	41	3.02	0.52	3.28	0.78	-2.561*

Table 5. Description of 6 novice college teachers' scores on IOC between the pre- and post-test

Teacher	N	Pre-test		Post-test		t
		Mean	SD	Mean	SD	
A	47	3.87	1.04	4.21	0.46	-3.061**
B	48	4.64	0.36	4.51	0.44	2.118*
C	28	3.75	0.37	3.60	0.86	1.213
D	46	3.91	0.62	4.25	0.62	-7.057***
E	51	3.94	0.56	3.96	0.70	-0.350
F	41	3.20	0.49	3.58	0.79	-4.322***

4.KSU (Knowledge of Students' Understanding)

Table 6 shows 6 novice teachers' change in the KSU dimension, that 4 (teachers A, D, E and F) of them achieved a level of significant difference from pre-test to post-test and their mean scores in the post-test are all higher than in the pre-test. Teacher C's post-test mean score is higher than in the pre-test, but does not achieve a level of significance. Teacher B's post-test mean score is lower than in the pre-test.

To further explore the reasons for the PCK change of the six new teachers and make a teaching diagnosis, as well as recommendations, the following section will discuss and compare the performances of the teachers through therelated qualitative auxiliary materials including: students' feedback, teachers' interview and teachers' reflection in the workshops and video recordings.

Table 6. Description of 6 novice college teachers' scores on IOC between the pre- and post-test

Teacher	N	Pre-test		Post-test		t
		Mean	SD	Mean	SD	
A	47	3.79	1.12	4.18	0.47	-3.182*
B	48	4.43	0.53	4.38	0.48	0.614
C	28	3.72	0.41	3.81	0.81	-0.619
D	46	3.84	0.64	4.24	0.58	-5.786***
E	51	3.74	0.77	3.88	0.69	-2.249*
F	41	3.17	0.51	3.59	0.83	-4.864***

1. Teacher A's assignment is heavy and examination content is too difficult; she discovers the teaching blind spots through reflection

For Teacher A of the "Chinese literature" course, some students provided the following feedbacks to the open-ended questions, such as:

Student 04: I have learned almost nothing, there are too many reports, and it is hard to understand the teacher's intention.

Student 08: There are too many unnecessary tests. I don't know what is the point of all those!

Student 12: The tests are too difficult, and it is impossible to assess the learning effects of students

[Class A –first survey]

From the students' feedbacks in the first survey, it can be seen that students generally think that the reports and evaluation methods of Teacher A cannot test their understanding of the subject contents. In addition, some students also said that the examinations and subject contents are rather difficult. Therefore, the teacher should specifically understand their learning difficulties, and modify course contents and evaluation methods to enhance the learning achievements of the students. Teacher A reflects constantly and discusses with other teachers when participating in the seminars, for example:

Teacher A: My biggest gain from the questionnaire is the open feedbacks of students, amounting to as many as 30. Most students do not clearly understand from the very beginning why they have to take the Chinese course. Hence, their attitudes toward the classes are very polarized as some students sitting at the back of classroom do not want to listen at all. They believe it is unnecessary.

Teacher B: You can communicate with students in the first class. I did not explain the textbook in great detail. Instead, I first communicated with them explaining why we should take the Chinese class in our freshman year.

[Workshop]

Taking advantage of the seminars to exchange views with other teachers, Teacher A found out that the teaching blind spot was actually the failure to notice the individual differences between students who were polarized in their learning attitude. Hence, it is important to design appropriate teaching contents and methods aimed at those students not motivated in learning.

2. Teacher B's teaching is animated and vivid; realizing from the reflection that theme-based teaching is most suitable.

Teacher B teaches "Introduction to Chinese Culture". Some of the students' feedbacks and suggestions given in the open-ended questions are as follows:

> Student 30: The teacher is serious in teaching and the materials he prepares are informative.
> [Class B –first survey]
> Student 01: The teacher is very serious; I like his class very much; the class contents are vivid and interesting.
> Student 25: The teacher demonstrates his teaching contents using many graphics and PowerPoint.
> [Class B –second survey]

Although there is no significant difference between the results of the two surveys for Teacher B, his average points are the highest among the six teachers ($M > 4.40$). Despite his rich teaching experience, Teacher B is teaching in this university for the first year, and he voluntarily participates in the seminars to improve his professional skills. The interviews show that Teacher B conducts his class in a lively and vigorous matter, and is able to use many graphics and PowerPoint presentations for illustration.

> Teacher B: In fact, this subject is very interesting. I conduct every class in a theme-based manner and ask students to do some homework of field investigation. For example, when teaching the topic of food vendors in Taiwan, I asked students to try out various Taiwanese foods, and take photos as evidence to write a report on this topic. When teaching the topic of china or artifacts, I asked them to go to the National Palace Museum, or at least view the objects on the Internet.
> [Interview]

3. Teacher C teaches too fast, so that students cannot fully understand the contents taught; realizing from the reflection that he is not aware of the students' prior knowledge.

Teacher C teaches "Introduction to Information System and Network". Some of students' feedbacks and suggestions provided in the open-ended questions are as follows:

> Student 07: Sometimes the teacher teaches too fast and even jumps to the next section even before we fully understand the content.
> Student 14: The implementation process is not sufficient due to the time limit; it would be perfect if we are not in such a hurry!
> Student 15: Some contents are relatively difficult; we hope the teacher can teach it twice!
> [Class C –first survey]

Generally, students think that he teaches too fast. Hence, the teacher should explain, illustrate and practice more about the units or parts that the students do not understand. During the workshops, Teacher C has reviewed this problem after seeing the videotape and pointed out that "*the majority of novice teachers tend to cause learning disruptions to students when teaching at inappropriate pace; hence, more attention should be paid in the future*".

Therefore, the teacher should more precisely understand the prior knowledge and learning difficulties of students as the basis for teaching design and evaluation. Regarding the contents that are relatively difficult, more illustrations or analogies should be given, or students should be encouraged to express their opinions in order to create a positive interaction between the teacher and students.

4. Teacher D's course content is difficult and the class is outnumbered; realizing from the reflection that appropriate IRS should be adopted.

Teacher D teaches "Introduction to Computer Science". Some of the students' feedbacks and suggestions provided in the open-ended questions are as follows:

Student 01: Programming language is so hard!

Student 14: The class size is too large, resulting in an adverse learning environment; I wonder if we could move to a bigger classroom to guarantee that each person has access to one computer.

[Class D –first survey]

As seen, Teacher D should strengthen the explanation of the contents taught, especially about programming language. As to contents that are difficult to some students, they should be encouraged to raise questions at anytime. In addition, whether the learning environment is appropriate should be noted to provide students with more practicing opportunities.

Student 35: When we ask questions, the teacher would explain the abstract knowledge with examples, helping us to understand.

Student 08: The teacher uses diversified teaching methods, and I have started to like this subject.

[Class D –second survey]

According to the t-test, Teacher D has achieved a significant difference in the results of the two surveys. Computer programming language is rather difficult to students, as indicated in the first survey; however, the second survey reveals that the teacher was able to illustrate abstract knowledge with examples and use diversified teaching strategies to help students better understand the concept of the subject.

5. Teacher E's teaching assessment and arrangements need to be improved; he adopts more illustration and elaboration after reflection.

Teacher E teaches "Probability and Statistics". Some of students' feedbacks and suggestions provided in the open-ended questions are as follows:

Student 08: The class is easy, but the examinations are rather hard with tricky questions.
Student 05: The teaching is too slow at the beginning; thus, it becomes very fast later; and I cannot keep up with the progress.
Student 18: The progress is too slow sometimes, and I get bored easily.

[Class E – second survey]
Researcher: It is suggested that Teacher E should make more illustrations and explanations on relatively difficult units or parts, and spend more time on those parts. Moreover, when designing the examinations, Teacher E has not clearly informed the students on the percentage of questions related to various topics in order for students to know how to prepare for examinations. [Workshop]

Students in this class expressed that the examinations were relatively hard and they did not know how to prepare for them. They expected that the teacher could lower the difficulty of the examinations. The teaching pace was relatively quick, and some students may have learning difficulties. Regarding arrangements for teaching content, Teacher E may modify the teaching progress according to the level of difficulty of the content (e.g. number of theme-based classes) so that students can better understand each theme. In addition, the teacher can encourage students to do more practice and calculation, and arrange quizzes or practices, in order for students to understand the problems and for the teacher to be aware of the students' learning difficulties.

6. Teacher F uses inappropriate teaching strategies; adjusting the teaching strategies through reflection.

Teacher F teaches "Engineering Mathematics". Some of students' feedbacks and suggestions provided in the open-ended questions are as follows:

Student 08: In terms of effectiveness in teaching this course, using PowerPoint is worse than using blackboard.
Student 22: It is hard to understand engineering mathematics illustrated by PPT.
Student 28: It is better to write on the blackboard when teaching engineering mathematics.
[Class F - first survey]

As seen, students generally complained that Teacher F's teaching strategy in using PPT for explanation and elaboration on calculation is inappropriate and leads to learning difficulties. Thus, Teacher F's average points in the second survey are lower than those of other teachers. Moreover, Teacher F claims that the students in that class are of different levels, and most of them are re-taking the course; thus, it is hard to teach them according to their levels. Teacher F keeps reflecting to search for proper solutions to the problem during the seminar courses.

Teacher F: Since many students are from the Department of Continuing Education or evening schools, and are taking the course for the second or even the third time, I am perplexed by their different levels! I really do not know how to teach the course. I think writing on the blackboard is too slow, and it is really more convenient to use the

computer in teaching! Turning pages and citing from here and there would be more convenient, really convenient!
[Interview]

In fact, PowerPoint could not be fully applied in the engineering course. Some calculation deduction processes would be much clearer when written on the blackboard. Hence, after taking the seminars and listening to the researcher's suggestions, Teacher F started to reflect on improving the teaching method, and using both PowerPoint and blackboard to cater for the students' thinking and learning models.

Researcher: One disadvantage of PowerPoint is that it presents all the answers and the procedures for reaching the answers. Therefore, thinking and interactivity would be weakened as the answers are already known! In fact, there is no fixed rule on using PowerPoint or the blackboard, they can be used together, and you may use PowerPoint to present graphs or summaries.
[Workshop]

DISCUSSION AND IMPLICATIONS

The main contribution of this study is the use of the survey to understand the overall teaching performance of the teachers and provide them with teaching reflection. Compared with a traditional semester-end evaluation, which could only collect rather few opinions, this design could also collect students' many opinions from open-ended questions and provide diagnostic functioning to allow new teachers to find out what to change or differences to make after a period of teaching, and make reflective thinking as well as timely modifications (Jang, Guan & Hsieh, 2009). By the mid-term survey, we could realize the students' perceptions of the teacher's PCK to provide suggestions on revising teaching techniques; on the other hand, by the final survey, we could further realize if the teacher's PCK was promoted or not after the reflection and revision. Moreover, the research results could make the participated teachers revise and design the courses next semester. From the students' comments, it has been found that the four dimensions of the questionnaire (SMK, IRS, IOC and KSU) should be taken into consideration regarding practical teaching. Teachers can learn about their performace and changes needed to be made from such dimensions. However, in fact, it symbolizes the overall change of teachers in terms of PCK for the paper length's shake.

With regard to the PCK of the teachers in this research, Teachers A and D showed good progress and PCK growth during the research period. In spite of the slightly lower points, Teacher B maintained a certain level of performance. Teachers C, E, and F needed further improvements and modifications in all four dimensions. In the first survey, some students complained that Teacher A gave too much homework and the examinations were rather difficult. After reflection, she realized her teaching blind spots and noticed the individual differences among the students. Students suggested that Teacher B's teaching was lively and vivid. Teacher B found that he is good at theme-based teaching, allowing students to explore the fun of life. Students complained that Teacher C taught too fast, and they could not fully understand the contents taught. After reflection, Teacher C found that he did not understand the prior knowledge and learning difficulties of the students. Students expressed that Teacher D's course was relatively hard and the class size too big. After reflection, Teacher D decided to adopt diversified teaching strategies and features to overcome the teaching difficulties. Students expressed that Teacher E's examinations were rather hard and the teaching pace is not consistent. After reflection, it was recommended to adopt more illustrations and explanations to allow students to better understand the teaching method. Finally, students complained that Teacher F's teaching strategies were not appropriately used. Before the midterm, students claimed that the teacher focused on slides when teaching in class. However, after the midterm, the teacher adjusted his teaching and added board writing. Reflection enables them to consider what the most effective teaching strategies would be.

In the study, the reflective model including students' feedback, workshop, instructional videotaping, and interviews were adopted to prove the changes of teacher's PCK and the action of reflection. As to the students' feedback about the progress of these courses, the workshop acted as a platform for teachers to interact with each other. The teachers should consider the teaching pace and adjust their teaching strategies to be the reference of later courses. In instructional videotaping, by way of micro-teaching or peer evaluation, it was easy to cause pressure for teachers. Consequently, we helped the teachers to record their instructions, transformed a DVD, and let them self-reflect. In this way, we aimed to provide the opportunities of individual teaching reflection instead of bringing into the consideration of instructional evaluation. It made the teachers face real teaching programs and tried to find the appropriate solutions (Clarke, 2009). At last, during the interviewing with the case teachers, the researchers would ask about their feelings after watching the DVD, and then brainstormed with each other to find ways to reform their teaching according to students' feedback and suggestions.

There is very little research on the development of a university teachers' PCK. Major and Palmer (2006) conducted a qualitative study on PCK using problem-based learning activities of the university teachers, and recommended that universities should hold study programs to help teachers evaluate PCK changes. In view of this study, six new teachers participated in the survey, and exchanged views with other teachers and the researchers in the PCK seminars to find out the solutions. Moreover, reflective teachers are involved in comparing the quality of their teaching (Hammersley-Fletcher & Orsmond, 2005). The feature of this study is to analyze, using both quantitative and qualitative methods, to make modifications in teaching. In the final workshop and interview, the researchers made an analysis on the questionnaire results, and gave suggestions to the six teachers.

The six new teachers reflected on their teaching during the PCK workshops, gradually constructing individual knowledge and a teaching method (Magnusson et al., 1999). The developed model provides a way to enhance university teachers' PCK and also serves as useful reference for other pedagogical goals. Future studies can focus on how to convert subject skills and discussion about methods, and integrate more qualitative data for analysis and illustration. The following research limitations and difficulties were encountered in the process of this study: 1) it is not easy to arrange the workshops since teachers have different teaching schedules, and not all the teachers can participate in the entire research process; 2) students' class attendance varies; thus, the number of samples in the two surveys cannot be accurately controlled, resulting in fewer valid samples; 3) the time of the questionnaire survey is determined by the teachers, but the students' willingness to fill out the questionnaire is significantly lower before examinations.

If the above limitations can be overcome, the implementation of a survey and the participation during a seminar may be more effective. Hence, future studies can use an online questionnaire method to more accurately track the number of respondents and students' feedback for a more in-depth analysis.

REFERENCES

Abell, S.K. (2008). PCK twenty years later: Does it remain a useful idea? *International Journal of Science Education, 30,* 1405–1416.

American Association for the Advancement of Science (AAAS). (1993). Benchmarks for scientific literacy. Washington, DC: Author.

Carlsen, W.S. (1993). Teacher knowledge and discourse control: Quantitative evidence from novice biology teachers' classrooms. *Journal of Research in Science Teaching, 30,* 471–481.

Carter, K. & Doyle, W. (1987). Teachers' knowledge structure and comprehension process. In J. Calder-head (Ed.), *Exploring teachers' thinking* (PP.147-160). London; Cassel.

Clarke, L. (2009). Video reflections in initial teacher education. *British Journal of Educational Technology, 40*(5), 959-961.

Clegg, S., Tan, J. & Saeidi, S. (2002) Reflecting or acting? Reflective practice and continuing professional development in higher education, *Reflective Practice, 3*(1), 131–146.

Dalgarno, N. & Colgan, L. (2007). Supporting novice elementary mathematics teachers' induction in professional communities and providing innovative forms of pedagogical content knowledge development through information and communication technology. *Teaching and Teacher Education, 23,* 1051-1065.

De Jong, O., Van Driel, J. H. & Verloop, N. (2005). Preservice teachers' pedagogical content knowledge of using particle models in teaching chemistry. *Journal of Research in Science Teaching, 42*(8), 947-964.

Erickson, F. (1986). Qualitative methods in research on teaching. *Handbooks of research on teaching.* 119-161. New York: Macmillan.

Fernandez-Balboa, J. & Stiehl, J.(1995). The generic nature pf pedagogical content knowledge among college professors. *Teaching and Teacher Education, 11*(3), 293-306.

Gess-Newsome, J. & Lederman, N.G. (1993). Preservice biological teachers' knowledge structures as a function of professional teacher education: A year-long assessment. *Science Teacher Education, 77*(1), 25-43.

Grossman, P.L. (1990). The making of a teacher: *Teacher knowledge and teacher education.* New York: Teachers College Press.

Hammersley-Fletcher, L. & Orsmond, P.(2005). Reflecting on reflective practices within peer observation. *Studies in Higher Education, 30*(2), 213–224.

Hashweh (2005) Teacher pedagogical constructions: a reconfiguration of pedagogical content knowledge. *Teachers and Teaching: Theory and Practice, 11*(3),-273-292.

Hasweh, M. (1987). Effects of subject matter knowledge in the teaching of biology and physics. *Teaching and Teacher Education*, *3*(1), 109-120.

Jang, S. J. (2009a). *PCK and teaching innovations*. NY: Nova Science Publishers.

Jang, S. J. (2009b). Development of a research-based model for enhancing PCK of secondary science teachers. In A. Selkirk & M. Tichenor (Ed.) *Teacher Education: Policy, Practice and Research* (PP.189-212). NY: Nova Science Publishers.

Jang, S. J., Guan, S. Y. & Hsieh, H.F. (2009). Developing an instrument for assessing college students' perceptions of teachers' pedagogical content knowledge. *Procedia Social and Behavioral Sciences, 1*(1), 596-606.

Knight, S. L. & Waxman, H. C. (1991). Analyzing effective teaching of Hispanic students' problem-solving strategies in Spanish. *NABE Annual Conference Journal, 1988-1989*. Washington, D.C.: National Association for Bilingual Education.

Lederman, N.G., Gess-Newsome, J. & Latz, M.S. (1994). The nature and development of preservice science teachers' conceptions of subject matter and pedagogy. *Journal of Research in Science Teaching, 31,* 129-146.

Lee, H.-J. (2005). Understanding and assessing preservice teachers' reflective thinking. *Teaching and Teacher Education, 21*, 699–715.

Lenze, L.F. & Dinham, S. M. (1994). Examining pedagogical content knowledge of college faculty new to teaching. Paper presented at the Annual Meeting of the American Educational Research Association, New Orleans, Louisiana.

Loughran, J. J. (2002). Effective reflective practice: In search of meaning in learning about teaching. *Journal of Teacher Education, 53*(1), 33–4.

Loughran, J.J., Mulhall, P. & Berry, A. (2004). In search of pedagogical content knowledge in science: Developing ways of articulating and documenting professional practice. *Journal of Research in Science Teaching, 41*, 370–391.

Magnusson, S., Krajcik, J. & Borko, H. (1999). Nature, sources, and development of pedagogical content knowledge. In J. Gess-Newsome & N.G. Lederman (Eds.), Examining pedagogical content knowledge (pp. 95–132). Dordrecht: Kluwer Academic.

Major, C. & Palmer, B. (2006). Reshaping teaching and learning: The transformation of faculty pedagogical content knowledge. *Higher Education, 51*(4), 619-647.

Mezirow, J. (1990). Fostering critical reflection in adulthood: A guide to transformative and emancipatory learning. San Francisco, CA, Jossey-Bass.

Mulholland, J. & Wallace, J. (2005). Growing the tree of teacher knowledge: Ten years of learning to teach elementary science. *Journal of Research in Science Teaching*, *42*(7), 767-790.

National Research Council (NRC). (1996). National science education standards. Washington, DC: National Academy Press.

Nilsson, P. (2008). Teaching for understanding: The complex nature of pedagogical content knowledge in pre-service education. *International Journal of Science Education*, *30*(10), 1281-1299.

Patton, M. Q. (1990). *Qualitative Evaluation and Research Methods*. London: Sage Publications.

Penny, A. R. & Coe R. (2004) Effectiveness of consultation on student ratings feedback: A meta-analysis, *Review of Educational Research. 74*(2), *215–253.*

Rhine, S. & Bryant, J. (2007). Enhancing pre-service teachers' reflective practice with digital video-based dialogue. *Reflective Practice, 8*(3), 345-358.

Sanders, L. R., Borko, H. & Lockard, J. D. (1993). Secondary science teachers' knowledge base when teaching science courses in and out of their area of certification. *Journal of Research in Science Teaching, 3,* 723-736.

Shulman, L. S. (1986). Those who understand: Knowledge growth in teaching. *Educational Researcher, 15*, 4-14.

Shulman, L. S. (1987). Knowledge and teaching: Foundations of the new reform. *Harvard Educational Review, 57*(1), 1-22.

Tobin & Garnett, (1988). Exemplary practice in science classroom. *Science Education*, 72, 197-208.

Tuan, H-L., Chang, H-P., Wang, K-H. & Treagust, D. F. (2000). The development of an instrument for assessing students' perceptions of teachers' knowledge. *International Journal of Science Education*, *22*(4), 385-398.

Van Dijk, E. M. & Kattmann, U. (2007). A research model for the study of science teachers' PCK and improving teacher education. *Teaching and Teacher Education, 23*(6), 885-897.

Van Driel, J. H., Beijaard, D. & Verloop, N. (2001). Professional development and reform in science education: The role of teachers' practical knowledge. *Journal of Research in Science Teaching*, *38*(2), 137-158.

Van Driel, J. H., Verloop, N., & De Vos, W. (1998). Developing science teachers' pedagogical content knowledge. *Journal of Research in Science Teaching, 35*, 673–695.

Van Driel, J.H., De Jong, O. &Verloop, N. (2002). The development of preservice chemistry teachers' PCK. *Science Education, 86*, 572-590.

Wallace, C.S. & Oliver, J.S. (2003). Journaling during a school-based secondary methods course: Exploring a route to teaching reflection. *Journal of Science and Teacher Education, 14*(3), 161–176.

Chapter 4

EVALUATING UNIVERSITY STUDENTS' PERCEPTIONS OF INSTRUCTORS' PCK USING STUDENT SURVEYS AND INSTRUCTOR WORKSHOPS

ABSTRACT

This research evaluates university students' perceptions of their instructors' PCK development using a developed instrument. There were 26 university instructors who participated in all three workshops and whose students received both the pre- and post-tests. PCK workshops were held for these university instructors to facilitate their self-reflection and the exchange of ideas among teachers. This study used a mixed method design, incorporating both quantitative and qualitative techniques. Data was collected, summarized and coded by surveys, interviews and PCK workshops. The results showed that more than half of the university instructors showed improvement in their PCK. Students considered that university instructors were strongest in Subject Matter Knowledge (SMK), and weaker in Instructional Representation and Strategies (IRS), and Knowledge of Students' Understanding (KSU). Furthermore, most instructors were able to modify their current teaching practices based on student mid-term feedback and thereby achieve positive results. The workshops provided instructors with opportunities for learning from each other and for reflecting on their teaching methods. Research implications of this study are provided along with suggestions for future research.

Keywords: PCK; Higher education; Professional development; Workshop intervention; College teaching and learning

INTRODUCTION

Some researchers have argued that longer-term professional development can provide opportunities for teachers at different professional stages to improve their knowledge of subject matter and their instructional strategies (Borko, 2004; Grossman Wineburg & Woolworth, 2001). Most university instructors entering the profession may find their initial teaching efforts stressful; but, with experience, they acquire a repertoire of teaching strategies and representations that they draw on throughout their teaching. While an instructor's teaching style provides a means of coping with many of the routine demands of teaching, there is also a danger that it can hinder his/her professional growth (Garcia & Roblin, 2008; Jang, 2011). Pedagogical Content Knowledge (PCK) is an important indicator for assessing the professional skills of university teachers (Jang, 2011; Lenze & Dinham, 1994). In fact, PCK has been described in many educational reform documents as a knowledge base necessary for effective teaching (American Association for the Advancement of Science, 1993; National Research Council, 1996).

In general, a survey to test the teaching effectiveness of university instructors is usually conducted at the end of each semester in order to determine the teaching performance of teachers during the previous period of time, and to help them make appropriate modifications. However, since most of the subjects in the current university curriculum are only taught for one semester, instructors do not have time to make modifications if student feedback is only made available at the end of the semester. In this way, they can only improve their teaching the next time they teach the course. To overcome the limitations on curriculum design and provide teachers with opportunities to make timely modifications, this questionnaire should be surveyed in the middle of the semester. This method can not only satisfy formative evaluation requirements, but also provide a diagnostic evaluation that can assist university instructors in improving their teaching effectiveness.

There have been two research streams in previous PCK studies. One research stream is related to elementary and secondary teachers' PCK, and another is associated with university teachers' PCK. Although there have been many studies to research and develop elementary and secondary teachers' PCK (Dalgarno & Colgan, 2007; De Jong, Van Driel & Verloop, 2005; Jang, 2009; Loughran, Mulhall & Berry, 2004), there has been much less attention given to university teachers' PCK. Furthermore, few studies have been completed on students' perceptions of teachers' PCK (Jang, 2011; Tuan, Chang, Wang & Treagust, 2000). We contend that, while student perception may differ from the teacher's self-perception, it provides a relatively objective account and an alternate view that supplements traditional teacher self-reporting. Another benefit of student-perceived instruments is that they make it easier to collect large samples for research. It is agreed that college students know enough to express their views of teachers' performance and that they are certainly key stakeholders in the educational field (Jang, Guan, & Hsieh, 2009). Their perception of an instructor's PCK should not only merit further investigation in this study, but also sheds more light on future PCK development. Therefore, this study evaluates university students' perceptions of instructors' PCK development using an instrument developed for this purpose. PCK workshops were also held for university instructors to facilitate the reflection and exchange of ideas among teachers.

THEORETICAL FRAMEWORK

Conceptualization and Components of PCK

A teacher's PCK is regarded not only as the fundamental knowledge and professional skills for teaching (Shulman, 1986, 1987), but also as the crucial factors that can be highly associated with teaching effectiveness and efficiency (Mansor, Halim & Osman, 2010; Segall,

2004). Comparing novice and experienced teachers' teaching performance, it has been shown that a novice teacher without enough teaching experience is very likely to have insufficient PCK, reducing their teaching performance. A recent study by Jang (2011) has further indicated that although some college teachers have a professional doctoral degree and sufficient subject matter knowledge, it is not easy for a college teacher without enough and satisfactory PCK to teach effectively and efficiently. According to Shulman (1987), PCK is described as "the blending of content and pedagogy into an understanding of how particular topics, problems, or issues are organized, represented, and adapted to the diverse interests and abilities of learners, and presented for instruction" (p.8). That is, PCK, which is "the category most likely to distinguish the understanding of the content specialist from that of the pedagogue" (Shulman, 1987, p.8), could be regarded as teachers' specific and professional knowledge that includes the combination and interaction of pedagogical and subject matter knowledge. Jang and Chen (2010) suggested that:

> PCK involves the transformation of other types of knowledge (subject matter knowledge, pedagogical knowledge, and knowledge of context) into viable instruction so that it can be used effectively and flexibly in the communication process between teachers and learners during classroom practice (p.554).

Magnusson et al. (1999) stressed that the development of PCK is determined by the content to be taught, the context in which the content is taught, and the way the teacher reflects on his/her teaching experiences. Although teaching experience was taken as the primary source of PCK development, reflection was also a critical component of professional development (De Jong, Van Driel & Verloop, 2005; Loughran, Mulhall & Berry, 2004). Therefore, reflection emerges as another important element for novice teachers in developing expertise in their practice and is central to their accepting more responsibility for their actions (Loughran, 2002; Shulman, 1987; Wallace & Oliver, 2003). The central goal of reflection is to develop teachers' understanding of why they employ certain instructional strategies and how they can improve their teaching to have a positive effect on students (Lee, 2005). A reflection process can be a development activity as well as a means to enhance an instructor's PCK. Reflective practice is becoming the favored paradigm for continuing professional development in higher education (Clegg, Tan & Saeidi, 2002).

Mezirow (1990) stated that reflection is an examination of the justification for one's beliefs, primarily to guide action and to reassess the efficacy of strategies and procedures used in the classroom. Reflective practice involves the process of teaching and the thinking behind it, rather than simply evaluating the teaching itself. Developing on this, Nilsson (2008) emphasized the role of teaching experience and reflection as a way of better understanding the complex entities that constitute a knowledge base for teaching. That study drew attention to the value of teachers participating in experiences that might contribute to the development of their PCK, and it supported the view of PCK development as a process of transformation.

Major and Palmer (2006) used a qualitative study of faculty members participating in a university campus-wide problem-based learning initiative to examine the process of transforming faculty PCK. They found that the existing knowledge of faculty and institutional intervention influenced new knowledge of faculty roles, student roles, disciplinary structures, and pedagogy. Teachers' PCK is deeply personal, highly contextualized and influenced by teaching interaction and experience (Van Driel, Beijaard & Verloop, 2001; De Jong, Van Driel & Verloop, 2005; Van Dijk & Kattmann, 2007).

Jang (2011) developed an instrument to evaluate college students' perceptions of the case physics teacher's PCK development. The survey was conducted twice in that study; with the pre-test carried out during the mid-term exam and the post-test, in the last week of the semester. The results showed that only SMK and IRS showed a significant difference in the four categories of this survey, while IOC and KSU did not.

The main advantage of using the university students' perception of a teacher's PCK is that that instructors tend to be influenced by self-righteous thinking since they are embodiments of authority in the classroom. Thus it is necessary to discuss this issue from the students' viewpoints to better understand the PCK development of teachers. In particular, because the students are involved in the teaching process, whether instructors have achieved the expected pedagogical objectives can be better understood through students' perception. Moreover, it is judged by all the students instead of a small number of observers.

Examining PCK through Student Perception

Penny and Coe (2004) pointed out that it has become a common feature in universities world-wide to employ student feedback as an indicator of teaching quality. Students' comments and perspectives can help instructors and researchers facilitate changes to the teaching, the teaching environment and the students' learning experience, as well as provide university teachers with appropriate support to enhance their teaching effectiveness.

Previous PCK studies have shown that there are two different perspectives to investigate the adequacy of a teacher's PCK. One perspective is to assess a teacher's PCK through their own perceptions (Johnston & Ahtee, 2006; Stump, 2001; Van Dijk, 2009). Another viewpoint can be obtained by examining a teacher's PCK through their students' common perceptions (Jang, Guan, & Hsieh 2009; Tuan et al., 2000).

Although several PCK studies have assessed the adequacy of teachers' PCK through pre-service or in-service teachers' perceptions, few studies have investigated PCK through the perceptions that the teacher's students have in common. Tuan et al. (2000) noted that "students' perceptions might not be consistent with the reality generated by outside observers, but they can present the range of reality for individual students and subgroups in the classroom" (p 386). It is clear that their viewpoints and insights should be very meaningful to practitioners and professionals in the educational field.

Moreover, with respect to the development of PCK components through university students' perceptions, a study by Jang, Guan, and Hsieh (2009) has further examined college students' perceptions of their teachers' PCK, finding that college teachers' PCK could be divided into four categories: "Subject Matter Knowledge (SMK), Instructional Representation and Strategies (IRS), Instructional Objects and Context (IOC), and Knowledge of Students' Understanding (KSU)" (p.596).

Accordingly, in view of the key influence of students' insights on teaching effectiveness and efficiency, accordingly, students' perceptions of their teachers' PCK are one of the focal points in this study.

Research Method

This study adopted a case study approach based on quantitative and qualitative analyses. A practical workshop and questionnaire survey were the two major tools in this study, and it is hoped that during the semester the new teachers can use the diagnosis and reflection to improve their instruction. Descriptive statistics were used to calculate the total means of the retrieved questionnaires from each class, as well as the means and standard deviations of each question, providing each teacher with the preliminary diagnostic results for their course and helping them to understand how their instruction is viewed by the students. In addition, the questionnaire includes an open question in which students can freely write about their difficulties in learning. The above feedback information forms a basis for teachers to improve their instruction.

Research Sample

Thirty-two new university teachers participated in the PCK Practical Workshop at one university. There were 26 teachers who participated in the complete three workshop series of courses and received both the pre- and post-tests, distributed to teachers in six different colleges: Humanities and Education (HE), Science (S), Engineering (E), Electrical Engineering and Computer Science (EECS), Design (D), and Business (B). Table 1 shows the college instructors (denoted by T1-T26 respectively), the number of students taking the course, and the number of valid samples collected before and after the survey.

Research Procedure

This study lasted for about 5 months, from the beginning of February 2011 to the end of June 2011, covering a full semester. The research model of university teaching (see Figure 1) was designed by the researcher and was comprised of three workshops, a mid-term assessment, a final assessment and interviews. Teaching reflection (center of the figure) was the main focus of this model and should continue throughout the study.

In the first workshop, the researcher described and discussed the study with the participating teachers. In addition, the researcher explaind related theories of PCK and the research tools in this study. For the mid-term assessment, the researcher used the questionnaires to survey students' perceptions of the instructors' PCK and then analyzed the results. During the second workshop, students' perceptions were analyzed through the results of the pre-test for instructors' PCK. Teachers with excellent teaching performances were invited to provide consultation with novice teachers who lack experience. Through interaction and discussion, the instructors could continue to reflect on the changes in various dimensions of PCK. Next, in the final assessment, the researcher used the questionnaires again as a post-test to evaluate any changes in the instructors' PCK after the teaching reflection. Then, in the third workshop, results of the pre-test and post-test questionnaires were compared and students' perceptions of instructors' PCK are discussed. Finally, the researcher interviewed the instructors and discussed the results from pre- and post-test questionnaire analyses.

Table 1. Description of the Colleges, Teachers, and Participants with Different Subject Majors

College	Teacher	Subject	Total	Valid Samples
HE	T1.	Introduction to Linguistics	62	36
	T2.	Professional Translation	36	34
	T3/	Practical English	36	32
	T4.	Chinese	63	38
S	T5.	General Physics Laboratory	60	32
	T6.	General Physics (2)	71	50
	T7.	General Physics (2)	72	52
	T8.	General Physics (2)	66	36
	T9.	Statistical Physics (2)	57	41
	T10.	Mathematics and culture	40	28
	T11.	General Physics (2)	40	35
E	T12.	Basic Concept of Computer Science	60	41
	T13.	Fluid Mechanics (1)	80	53
	T14.	Hydrology study of water	60	47
	T15.	Electric Circuits	88	64
EECS	T16.	Digital Logic Design	58	36
	T17.	Signals and Systems	60	42
	T18.	Design of Experiment	28	25
	T19.	Introduction to Computer Science (2)	65	45
	T20.	Introduction to Computer Science (2)	65	52
D	T21.	Historic Environment Regeneration History	34	26
	T22.	Landscape Design (2)	60	33
B	T23.	Economics (2)	78	42
	T24.	Financial Management	30	23
	T25.	NET Programming	60	38
	T26.	Statistics(2)	90	49

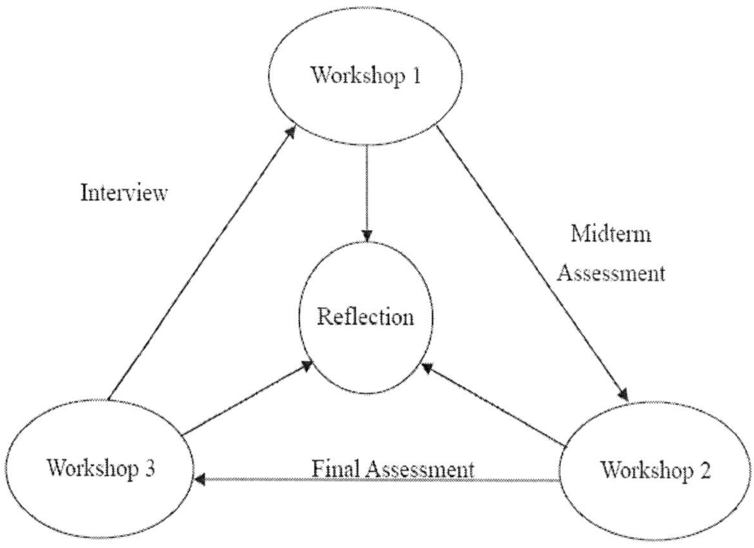

Figure 1. The research model of university teaching.

Data Collection and Tools

1. PCK Questionnaire

This study used the "Questionnaire on university students' perception of teachers' PCK" designed and developed by Jang, Guan and Hsieh (2009). The contents include the four dimensions of Subject Matter Knowledge (SMK), Instructional Representation and Strategies (IRS), Instructional Objects and Context (IOC), and Knowledge of Students' Understanding (KSU). There are seven sub-topics for each dimension, making a total of 28 items. The pilot study analyzed the questionnaire results from 182 college students, with a total of 172 valid samples collected. The questionnaire presented very high validity and reliability. Reliability was analyzed by internal consistency reliability, and analysis results showed a Cronbach's α value of 0.965, indicating good internal consistency of the 28 items. A validity analysis was conducted using a review by experts in PCK-related fields, and a factor analysis for verification and modification. In this study, the first and second surveys were conducted at the middle and at the end of the semester to understand the changes in students' perception of their teachers' PCK.

2. PCK workshops

The PCK Practical Workshops were held on March 23, May 4, and June 22 of 2011. The researcher acted as the main lecturer, and the lecture content progressed from easy to difficult. PCK workshops were mainly used to gather new university instructors from various colleges to join with educational research experts for discussion and exchange of views to help improve teaching through peer group stimulation and exchanging views. Teachers with excellent teaching performances and educational experts were invited during the seminar period to provide teaching consultation for reference by novice teachers with less teaching experience. Through interaction and discussion among teachers, these instructors could continue to reflect on the changes in various dimensions of their PCK. This enables the teachers to modify their teaching methods more quickly and effectively, based on what they have learned about students' learning needs in these courses.

3. Individual instructor interviews

The interview used survey questions designed to gain a deeper understanding of the students' conceptions of their teachers' PCK. According to the information gathered from the interviews, the researcher wished to confirm the validity of student responses to the two surveys. The teaching diagnosis report was produced according to the evaluation of teaching effectiveness and the analysis of integrated results of questionnaire. The results of the first survey were given to instructors after the mid-term, and the instructors had "reflections" after reading the report in the middle of the semester to enter into the "change" and "action" stage, in order to improve teaching effectiveness. The results of the second survey were given to instructors after the final examination to have a general view of the changes in their PCK as perceived by the students.

Data Analysis

This study used a mixed method design, incorporating both quantitative and qualitative techniques. Data was collected, summarized and coded by surveys, interviews and PCK workshops. The quantitative data includes the results from two surveys. The survey data was measured on a Likert 5-point scale, rated from 1 to 5 points according to the degree of difference. Finally, a comparative analysis of the two surveys was conducted by a t-test. The qualitative data includes the feedbacks of students and instructors, as well as a discussion on other qualitative data such as workshops. A qualitative analysis was carried out according to Patton (1990): 1) gather all the original data; 2) organize, categorize and edit the original data into files that can be easily identified and acquired; 3) summarize and identify important indexes for in-depth analysis according to study problems and types. In other words, data analysis is used to interpret and explain this data (Erickson, 1986). The triangulation method was employed to interpret the changes in PCK professional growth of the instructors. Then, the data was categorized and coded according to different interview content to establish specific data login categories.

RESEARCH RESULTS

1. Most of the University Instructors Showed Growth in PCK

Table 2 shows the pre-test and post-test mean scores as well as the standard deviation for teachers' PCK scores in the six colleges of this university. At total of 26 university instructors participated in this study, and 17 instructors had higher post-test means than pre-test means, showing that they developed after one semester, and five instructors reached a level of significance (T2, T3, T19, T20, and T26).

In terms of the different colleges, two of the four HE College teachers showed growth in post-test mean values; four of the seven College of Science teachers showed growth in post-test mean values; five of the seven College of Engineering teachers showed growth in post-test mean values, both of the two EECS College teachers showed growth in post-test mean values, one of the four College of Business teachers showed growth in post-test mean values, and both of the College of Design teachers showed growth in post-test mean values.

2. Students Consider That University Instructors Were Strongest in SMK, and Weaker in IRS and SKU

Table 3 shows the pre-test and post-test means for the four aspects of the PCK of instructors at different colleges. In all six colleges the students believed that SMK was the strongest aspect of these university instructors; whereas the HE, E, B and D College instructors had weaker IRS, and S and EECS College instructors had weaker KSU.

In the workshop discussions, most teachers agreed with the above views. They believed that university teachers were best at SMK, while they had to learn more in IRS and SKU, and hold a reflective and contemplative mood toward student feedback. For example, teacher T2 stated:

> From the midterm test, I found that the testing results showed SMK as the strongest part, while my scores in "instructional representations" were lower, with the lowest scores for items 11 and 12. Since my course is more theoretical, it is difficult to improve upon Item 12 ("My teacher uses demonstrations to help explain the main concepts."), so

in the post-test I was still lower in Item 12. However, regarding Item 11 ("My teacher provides opportunities for me to express my views during class."), after the midterm, as I prepared for lessons, I would make the PowerPoint presentation more detailed for the difficult concepts. I also lecture on them twice in class and a little more slowly, and I offer timely questions. Thus in the post-test, my score in Item 11 was higher. (Workshop)

Table 2. Performance of six College Teachers Pre & Post

Teacher	$Mean_1$ SD_1	$Mean_2$ SD_2	t
T1	4.347 .372	4.494 .647	-.353
T2	3.816 .451	3.967 .454	-2.156*
T3	3.275 .485	4.099 .747	-3.06*
T4	3.478 .410	3.362 .683	1.058
T5	4.110 .436	4.101 1.084	.026
T6	3.891 .630	3.922 .550	-.345
T7	3.206 .557	3.142 .673	.500
T8	3.921 .579	4.400 .617	-1.414
T9	4.497 .502	4.464 .362	.357
T10	4.350 .386	4.418 .321	-.561
T11	3.500 .695	3.518 .583	-.184
T12	4.053 .560	4.214 .248	-1.174
T13	4.025 .563	4.113 .589	-1.036
T14	3.422 .765	3.716 .748	.201
T15	4.086 .474	4.136 .467	-.795
T16	3.204 .442	3.402 .635	-.932
T17	3.615 .411	3.583 .382	.201
T18	4.120 .511	4.100 .395	.507
T19	2.891 .912	3.515 .592	-3.284*
T20	3.285 .467	3.500 .522	-2.24*
T21	3.287 .771	3.480 .810	-1.219
T22	4.629 .413	4.669 .253	-.254
T23	4.237 .378	4.177 .510	.507
T24	4.102 .529	4.009 .626	.849
T25	3.701 .849	3.952 .571	-1.686
T26	3.638 .552	3.923 .616	-2.523*

$^*p<.05$.

Furthermore, college students perceived that T14's use of multimedia such as PowerPoint is a good means to clearly present the teaching content. Nonetheless, the presentation seemed to be too fast. The following were the opinions from some students of this teacher from the pre-test:

S1: PowerPoint is a good tool to clearly present the teaching content. But the time to present the slides is too short.

S2: I think the instructor can use the blackboard to demonstrate the calculations.

T14: Both PowerPoint and the blackboard are evenly applied in my class. I use PowerPoint to clearly present the content that they've learned in high school. Then I use the blackboard to demonstrate the calculations step by step. In my opinion, it is more appropriate to introduce the formula and concepts by PowerPoint; on the other hand, it is more understandable to deduce the formula by writing on the blackboard. (Interview)

Table 3. Performance of six Colleges in four Dimensions: SMK, IRS, IOC and KSU

Dimension /College	SMK		IRS		IOC		KSU	
	$Mean_1$	$Mean_2$	$Mean_1$	$Mean_2$	$Mean_1$	$Mean_2$	$Mean_1$	$Mean_2$
HE	4.019	4.021	3.715	3.746	3.716	3.763	3.736	3.749
S	4.206	4.228	3.791	3.962	3.822	3.990	3.770	3.997
E	3.668	3.753	3.236	3.462	3.390	3.656	3.399	3.611
EECS	3.927	4.141	3.703	3.946	3.842	3.898	3.643	3.880
D	4.209	4.170	3.769	3.887	3.775	4.000	3.929	4.019
B	4.182	4.318	3.873	3.986	3.904	4.041	3.875	3.955

3. Most Instructors Were Able to Modify Their Current Teaching Practices Based on Student Midterm Feedback and Achieve Positive Results

Students' feedback of opinions helped the participating instructors to make some improvements in their instruction based on the questionnaire results, including an immediate adjustment of issues such as instructional method and content, speed of class presentation, and excessive homework. Student suggestions are very helpful for instruction, allowing instructors to improve upon problems or shortcomings in instruction. The following are some teachers' opinions:

T1: The midterm test was very helpful, and I was able to immediately adjust my instructional methods and content. (Workshop)

T5: I received approval of my teaching from the students. Following the student opinions that there was too much homework, I tried to make adjustments and to communicate with the students. (Workshop)

T10: Before the midterm test, many students gave negative feedback on this course. After seeing the results of the midterm test, I slowed my teaching pace, and focused on the main points. For class each week, I designed practice questions and asked students to do the exercises. Thus, after the mid-term, I found that there was more interaction between the teacher and students. But since this course did not involve technology, the technology portion did not improve. (Interview)

T12: I was happy to see student feedback at the end of the term showing that they are better able to understand and accept my teaching methods. However, it is still impossible to satisfy the needs of all students. Of course, I can still improve my instruction techniques. (Interview)

To show how students' opinions affected the instructors' teaching, we chose the "General Physics" course as a descriptive example. The students' opinions from the open-ended question of the survey for T8 are shown:

S3: The teaching pace is slightly fast.

S4: I'm not good at English, so it's a pain for me to have most of the course lectures in English.

S5: The teaching is good. But his teaching would be better if the teaching pace could be slower.

The midterm questionnaire was helpful for the instructor to evaluate his own teaching. T8 had referred to the midterm questionnaire and made some changes to his teaching, such as making his teaching pace slower after the midterm. Therefore, the student feedback was helpful for the instructor to improve his teaching. Examples from the interview data are as follows:

> Student feedback from the pretest shows they think my teaching pace is a little too fast. This might be because I am under pressure to finish the first seven chapters before the midterm; whereas after the midterm, there are only four chapters remaining. I have also compared my teaching pace with other instructors, and I find that I have to maintain this teaching pace to follow the syllabus. But on the other hand, I think the way I use PowerPoint to present my lecture might also be a reason that makes leading my teaching pace seem too fast. (T8 Interview)

In addition, this study finds that the midterm test can especially induce college students to express their opinions because they still hope their teachers would change teaching methods. The teaching practices of some teachers also gained student feedback. In feedback after the post-test, there was an increase in positive comments. Below are the open opinions from a student of T19 in the pre-test and post-test.

In the pre-test, the student responded as follows:

- It was pretty well taught.
- Learning is hard, would it be possible to talk slower or explain it more clearly, use more example questions and answer the questions.
- Some of the homework was too hard.
- I hope that the teacher would interact with us more, and explain the meaning of example questions and how to solve them. Often I only know how to memorize the formulas but do not know how to use them.
- In the example questions, the teacher should explain them more clearly, so we can understand how to make the calculations.
- Can you explain the main points and the evaluation standards?
- Sometimes the teacher talks too fast and some things are not clearly explained.

In the pos-test, the student responded as follows:

- Good, the lecture was very detailed.
- The speaking speed has been adjusted.
- Good.
- Thank you.

4. The Workshops Provide Instructors with Opportunities for Learning from Each Other and Reflection

Most teachers have their own blind spots in terms of their teaching methods, and this workshop can be used to evaluate one's own instruction, subject characteristics, number of students, and teaching methods, all of which can affect the effectiveness of instruction. This workshop was intended for teachers to share their instructional experiences and ask questions, in turn receiving peer feedback and suggestions, so as to learn from each other. In addition, the workshop provides teachers with opportunities to reflect and understand the students' pre-existing knowledge.

This study found that the workshops can help teachers grasp all aspects of instruction, to understand, transform, evaluate, reflect, and produce a new understanding. This process can make instruction a more enjoyable enterprise, with a sense of accomplishment. In the art of teaching if there can be more design in instruction, performance activities and enthusiasm can be incorporated, instruction can continue to develop. Below are some teacher responses.

T5: The prior knowledge of students is important in my course. However, when students take this course, they have already forgotten what they learned before. Therefore, after the midterm, I spent about one class to review the pre-existing knowledge to help student learning, and this helps their research. (Interview)

T2: After so much discussion, I believe that the most important thing in instruction is enthusiasm. This workshop can help teachers share and encourage each other, and make us even better. (Workshop)

In the workshop discussion, T21 posed the following question and received enthusiastic responses from other teachers:

T21: For this class, I tried many instructional strategies. But even though the test scores at the end of the term increased, the overall learning atmosphere in class still did not improve. Some students seem to be only concerned with their own affairs. They are not absent, but there is no sense of participation in class. I wanted to ask how other teachers deal with this type of student?

T19: In the past, I tried to use individual counseling sessions to help students that are frequently absent, and after these sessions I found that their attendance rates increased.

T20: I found that peer pressure can enhance the cohesion of this class.
I first divided them in groups, and every time I had class, if everyone in a group shows up, they can add two points to their final grade; in this way, the attendance rates are as high as 90%.

T12: I would buy candy and very spicy chewing gum, and let students who are falling asleep in class wash their face; then would give them the candy.

T23: In the group discussion method, I let the groups discuss, and after they discuss, they can write the answers on the blackboards; whoever is fastest and right can have extra points.
At the end of the semester, students commented that they realized economics could be interesting after all.

Researcher: Most teachers only focus on overall differences, and Teacher T21 could note the small differences, especially that some students could influence the overall learning atmosphere – this is very good. (Workshop)

DISCUSSION AND IMPLICATIONS

In addition to continuing the last in-depth study on university physics teachers (Jang, 2011), this study expands this developmental tool and the workshop model to add more new teachers from different departments and colleges at this university. Other than a major increase in the number of participating teachers, it is of great urgency to enhance the PCK of university instructors. This study found that over half the university instructors in this program improved their PCK. It is believed that student feedback opinions helped participating teachers improve their instruction, including the immediate adjustment of instructional methods and content, teaching pace, and amount of homework. Furthermore, this study finds that the midterm test can effectively elicit student opinions on instructional methods since they have the hope that teachers can modify instructional methods to better conform to student needs. Compared with a traditional semester-end evaluation, which collects relatively few opinions (Jang, Guan & Hsieh, 2009), this design could also collect opinions from many students by using open-ended questions. This can have a diagnostic function, allowing new teachers to find ways they could change after a period of teaching, and support reflective thinking as well as make timely modifications.

With regard to the PCK of instructors in this research, students considered that university instructors were strongest in SMK, and weaker in IRS and SKU. Most new college teachers with a doctoral degree and a certain level of knowledge on their subject matter knowledge cannot teach effectively (Abell, et al., 2009; Major & Palmer, 2006). One of the reasons was that, unlike primary and secondary school teachers, they did not need to get a teacher's certificate (Jang, 2011). Second, aside from the subject matter knowledge, college teachers needed more pedagogical knowledge (Lenze & Dinham, 1994). In this study, there were three factors to support teacher instruction improvement. The first is student feedback data from questionnaire surveys, including both qualitative and quantitative answers. The second is the educational knowledge and peer opinion exchange in the course. Finally, the third is the teacher's own reflections and accumulated instructional experience. To evaluate the first two factors, this study used a questionnaire design and held workshops. New teachers responded positively to the former, helping to improve their instruction techniques. The workshop showed that they hoped to have more time and opportunities to interact among their peers, to pose questions and discuss the teaching diagnostics results, and engage in a self-evaluation of the questionnaire results. Thus, the promotion of teaching reflection became the main factor for short-term improvement in instruction by teachers.

The functions provided by the practical workshop included professional development in PCK for the subjects, understanding the importance of teacher evaluations, experiencing effects and practical implementation. The workshops were also an indispensable space for teachers to exchange experiences and opinions. In this study, the workshop provided teachers with opportunities to learn from each other and to reflect. Each teacher had his or her own blind spots, and this workshop can be used to evaluate one's own instruction, share instructional experiences, and in turn receive peer feedback and suggestions, so as to learn from each other. In addition, the workshop provided teachers with opportunities for reflection, to understand the preexisting knowledge and learning difficulties of students, and in turn adjust their own instructional content, strategies, and features. Most instructors at the workshop responded that the midterm evaluation was very beneficial.

Since university instructors have diverse courses, and each person has his or her own specialty in teaching, it is difficult to achieve a ceiling-style evaluation to establish teacher instruction evaluation systems. Thus, it is suggested that the case study method be used, focusing on the case study of personal development for college teachers, and it is not recommended to generalize on departments, colleges, or universities. It is hoped that teacher development may depart from the framework of departments and colleges, and return to the beginning point of personal professional development. This can assist teachers to increase their metacognitive experience in understanding their own strengths and weaknesses in teaching and increase their ability in personal evaluations. The teacher development departments at universities should provide assistance to monitor the personal instructional development of teachers; this would provide a channel for assistance to new instructors as they begin their teaching careers.

The main contribution of this current study is the use of a survey to understand the overall teaching performance of the teachers and provide them with teaching reflection. Here, 26 new teachers participated in the survey, and exchanged views with other teachers and the researcher in PCK seminars to look for solutions. Moreover, reflective teachers are involved in comparing the quality of their teaching (Hammersley-Fletcher & Orsmond, 2005). An important feature of this current study is to analyze using both quantitative and qualitative methods to make modifications in teaching. In the final seminar study and discussion, the researcher analyzed the questionnaire results and gave suggestions to the novice teachers.

New perspectives in further research are concerned with the implementation of the questionnaire of this study and if the instrument is suitable for different universities. After analyzing the survey data collected from university students' feedback in class, as well as the follow-up testing of reliability and validity of the collected questionnaire, a college teachers' self-examination mechanism was established for teaching performance and efficiency related to PCK. Future studies can focus on how to convert subject skills and discussion about methods, and how to integrate more qualitative data for analysis and illustration. The following research limitations and difficulties were encountered in this study: 1) It was difficult to arrange the seminars since teachers have different teaching schedules, and not all the teachers can participate in the entire research process; 2) Since students' class attendance varies, the number of samples in the two surveys cannot be accurately controlled, resulting in fewer valid samples; 3) the time of questionnaire survey was determined by teachers, but the students' willingness to fill out the questionnaire was significantly lower before examinations.

If these limitations can be overcome, the survey implementation and seminar participation may be improved. Hence, future studies can use an online questionnaire method to more accurately track the number of respondents and the students' feedback for a more in-depth analysis.

REFERENCES

Abell, S., Rogers, M., Hanuscin, D., Lee, M. & Gagnon, M. (2009). Preparing the next generation of science teacher educators: A model for developing PCK for teaching science teachers. *Journal of Science Teacher Education, 20*(1), 77-93.

American Association for the Advancement of Science (AAAS). (1993). Benchmarks for scientific literacy. Washington, DC: Author.

Borko, H. (2004). Professional development and teacher learning: Mapping the terrain. *Educational Researcher, 33*(8), 3 – 15.

Clegg, S., Tan, J. & Saeidi, S. (2002) Reflecting or acting? Reflective practice and continuing professional development in higher education, *Reflective Practice, 3*(1), 131–146.

Dalgarno, N. & Colgan, L. (2007). Supporting novice elementary mathematics teachers' induction in professional communities and providing innovative forms of pedagogical content knowledge development through information and communication technology. *Teaching and Teacher Education, 23*(7), 1051-1065.

De Jong, O., Van Driel, J. H. & Verloop, N. (2005). Preservice teachers' pedagogical content knowledge of using particle models in teaching chemistry. *Journal of Research in Science Teaching, 42*(8), 947-964.

Erickson, F. (1986). Qualitative methods in research on teaching. *Handbooks of research on teaching*. 119-161. New York: Macmillan.

Garcia, L. M. & Roblin, N. P. (2008). Innovation, research and professional development in higher education: Learning from our own experience. *Teaching and Teacher Education, 24*(1), 104-116.

Grossman, P., Wineburg, S. & Woolworth, S. (2001). Toward a theory of teacher community. *Teachers College Record, 103*(6), 942 – 1012.

Hammersley-Fletcher, L. & Orsmond, P. (2005). Reflecting on reflective practices within peer observation. *Studies in Higher Education, 30*(2), 213–224.

Jang, S. J. (2009). *PCK and teaching innovations*. NY: Nova Science Publishers.

Jang, S. J. (2011). Assessing college students' perceptions of a case teacher's pedagogical content knowledge using a newly developed instrument. *Higher Education, 61*(6), 663-678.

Jang, S. J. & Chen, K. C. (2010). From PCK to TPACK: Developing a transformative model for pre-service science teachers. *Journal of Science Education and Technology, 19*(6), 553-564.

Jang, S-J., Guan, S-Y. & Hsieh, H-F. (2009). Developing an instrument for assessing college students' perceptions of teachers' pedagogical content knowledge. *Procedia Social and Behavioral Sciences, 1*(1), 596-606.

Johnston, J. & Ahtee, M. (2006). Comparing primary student teachers' attitudes, subject knowledge and pedagogical content knowledge needs in a physics activity. *Teaching and Teacher Education, 22*(4), 503-512.

Lee, H.-J. (2005). Understanding and assessing preservice teachers' reflective thinking. *Teaching and Teacher Education, 21*, 699–715.

Lenze, L.F. & Dinham, S. M. (1994). Examining pedagogical content knowledge of college faculty new to teaching. Paper presented at the Annual Meeting of the American Educational Research Association, New Orleans, Louisiana.

Loughran, J.J. (2002). Effective reflective practice: In search of meaning in learning about teaching. *Journal of Teacher Education, 53*(1), 33–4.

Loughran, J.J., Mulhall, P. & Berry,A. (2004). In search of pedagogical content knowledge in science: Developing ways of articulating and documenting professional practice. *Journal of Research in Science Teaching, 41*, 370–391.

Magnusson, S., Krajcik, J. & Borko, H. (1999). Nature, sources, and development of pedagogical content knowledge. In J. Gess-Newsome & N.G. Lederman (Eds.), Examining pedagogical content knowledge (pp. 95–132). Dordrecht: Kluwer Academic.

Major, C. & Palmer, B. (2006). Reshaping teaching and learning: The transformation of faculty pedagogical content knowledge. *Higher Education, 51*(4), 619-647.

Mansor, R., Halim, L. & Osman, K. (2010). Teachers' knowledge that promote students' conceptual understanding. *Procedia Social and Behavioral Sciences, 9*, 1835-1839.

Mezirow, J. (1990). Fostering critical reflection in adulthood: A guide to transformative and emancipatory learning. San Francisco, CA, Jossey-Bass.

National Research Council (NRC). (1996). National science education standards. Washington, DC: National Academy Press.

Nilsson, P. (2008). Teaching for understanding: The complex nature of pedagogical content knowledge in pre-service education. *International Journal of Science Education, 30*(10), 1281-1299.

Patton, M. Q. (1990). *Qualitative Evaluation and Research Methods*. London: Sage Publications.

Penny, A. R. & Coe R. (2004) Effectiveness of consultation on student ratings feedback: A meta-analysis, *Review of Educational Research. 74*(2), *215–253.*

Segall, A. (2004). Revisiting pedagogical content knowledge: the pedagogy of content/the content of pedagogy. *Teaching and Teacher Education, 20*(5), 489-504.

Shulman, L. S. (1986). Those who understand: Knowledge growth in teaching. *Educational Researcher, 15*, 4-14.

Shulman, L. S. (1987). Knowledge and teaching: Foundations of the new reform. *Harvard Educational Review, 57*(1), 1-22.

Stump, S. L. (2001). Developing pre-service teachers' pedagogical content knowledge of slope. *The Journal of Mathematical Behavior, 20*(2), 207-227.

Tuan, H-L., Chang, H-P., Wang, K-H. & Treagust, D. F. (2000). The development of an instrument for assessing students' perceptions of teachers' knowledge. *International Journal of Science Education, 22*(4), 385-398.

Van Dijk, E. M. & Kattmann, U. (2007). A research model for the study of science teachers' PCK and improving teacher education. *Teaching and Teacher Education, 23*(6), 885-897.

Van Dijk, E. M. (2009). Teachers' views on understanding evolutionary theory: A PCK-study in the framework of the ERTE-model. *Teaching and Teacher Education, 25*(2), 259-267.

Van Driel, J. H., Beijaard, D. & Verloop, N. (2001). Professional development and reform in science education: The role of teachers' practical knowledge. *Journal of Research in Science Teaching, 38*(2), 137-158.

Van Driel, J.H., De Jong, O. &Verloop, N. (2002). The development of preservice chemistry teachers' PCK. *Science Education, 86*, 572-590.

Wallace, C.S. & Oliver, J.S. (2003). Journaling during a school-based secondary methods course: Exploring a route to teaching reflection. *Journal of Science and Teacher Education, 14*(3), 161–176.

Chapter 5

Assessing University Students' Perceptions of Their Physics Instructor's TPACK Using a Newly Developed Instrument

Abstract

Although there has been a plethora of studies investigating pre-service and in-service teachers' TPACK, few studies have been completed on college students' perceptions of teachers' TPACK. The primary purpose of this study was to assess the university students' perception of the case instructor's TPACK development. A physics instructor and his total class of 70 students participated in this study. This study used a mixed method design, incorporating both quantitative and qualitative techniques. A developed questionnaire was distributed to the students at the middle and the end of the semester, respectively. The study adopted constant comparison and a triangulation method related to the instructor's interviews, students' opinions and quantitative data to explain the instructor's TPACK change. The finding showed that the case Physics teacher's Technology Integration and Application (TIA) score significantly increased from the middle to the end of the semester; however, no significant changes in Subject Matter Knowledge (SMK), Instructional Representation and Strategies (IRS), and Knowledge of Students' Understanding (KSU) scores were detected. This study provided valuable insight and details related to the development of the case teacher's TPACK and provided program feedback on effective approaches in encouraging this development. The research implications of this study are provided along with suggestions.

Keywords: College teaching and learning; Students' perception; TPACK; Professional development

1. Introduction

Most university instructors entering the profession may find their initial teaching efforts stressful, but with experience, they acquire a repertoire of teaching strategies and representations that they draw on throughout their teaching career. While an instructor's style of teaching provides a means of coping with many of the routine demands of teaching, there is also a danger it can hinder his/her professional growth (Jang, 2011). In today's world, it cannot be denied educational technology has gradually played a critical role in our lives, probably because applications of technology to learning environments could be closely linked to students' learning achievement (Kopcha, 2010). Angeli and Valanides (2009) suggested that if teachers could make good and proper use of the information and communication technology (ICT), it is possible they could create better learning environments for students. Nevertheless, successful incorporations of ICT into instruction are not that simple. Unlike merely adding a new technology to classroom instructions, applying ICT to classrooms requires teachers to simultaneously have sufficient pedagogical content knowledge (PCK) and technological knowledge to maximize teaching effectiveness and efficiency (Jang, 2010). In line with this, numerous studies have focused on the issues of Technological Pedagogical and Content Knowledge (TPACK) to help teachers achieve more positive and preferable teaching and learning. The TPACK proposed by Mishra and Koehler (2006) has been one of the focal points in recent educational technology studies (Archambault & Barnett, 2010; Jang, 2010; Jimoyiann, 2010; Kramarski & Michalsky, 2010). Generally, there have been two research streams in previous TPACK studies. One research stream is related to pre-service and in-service teachers' TPACK, and the oother is associated with students' perceptions of teachers' TPACK. Although there has been a plethora of studies investigating pre-service and in-service teachers' TPACK (Angeli & Valanides, 2009; Chai, Koh, & Tsai, 2010; Kramarski & Michalsky, 2010; Niess, 2005), few studies have been completed on students' perceptions of teachers' TPACK (Jang & Chen, 2010a; Tuan, Chang, Wang & Treagust, 2000). We contend, while student perception may not be the same as teacher's self-perception, it provides a relatively objective account and alternative view to supplement traditional teacher self-reports. Another benefit of student-perceived instruments is that they make it easier to collect large samples for research. It is agreed that college students know enough to express their views of teachers' performance and key stakeholders in the educational field (Jang, 2011). Their perception of teachers' TPACK should not only merit further investigation in this study, but can also shed more light on future TPACK development. Therefore, further understanding the TPACK structure through different lenses is one of the focal points of this study. In addition, it has been shown that previous studies have highly focused on the theoretical development of TPACK, mainly due to the complex and unclear framework (Angeli & Valanides, 2009; Archambault & Barnett, 2010; Chai et al., 2011; Jimoyiann, 2010). Considering the key influence of this research gap on students' perception of teachers' TPACK, and the significant role of TPACK in teaching and learning effectiveness, the main purpose of this study was to assess the university students' perception of the case instructor's TPACK development between the middle and end of the semester.

2. Literature Review

2.1. TPACK Theoretical Framework

The conceptualization of TPACK is mainly derived from the theoretical framework of Pedagogical Content Knowledge (PCK). It has been shown that a teacher's PCK, containing at least pedagogical knowledge and subject matter knowledge, could be highly associated with their teaching effectiveness and efficiency (Shulman, 1986, 1987). Niess (2005) indicated that a teacher's PCK is an integration and interaction of teaching knowledge and subject matter knowledge. Jang and Chen (2010b) added:

> PCK involves the transformation of other types of knowledge (subject matter knowledge, pedagogical knowledge, and knowledge of context) into viable instruction so that it can be used effectively and flexibly in the communication process between teachers and learners during classroom practice (p. 554).

In other words, PCK not only stresses the important role of a teacher'spedagogical knowledge and subject matter knowledge during instruction, but also emphasizes their understanding of student learning capabilities and obstacles (Angeli & Valanides, 2009). Although there is no accurate and precise definition of PCK, generally, it is agreed the content of PCK could broadly include "teachers' knowledge of representations of subject matter, and their knowledge of learners' conceptions and content-related difficulties" (Angeli & Valanides, 2009, p.156).

The TPACK framework was built upon Shulman's PCK to include technology knowledge as situated within content and pedagogical knowledge (Mishra & Koehler, 2006). It has been shown that teachers' technological knowledge is another critical component that could be highly associated with teaching effectiveness and efficiency (Angeli & Valanides, 2009; Archambault & Barnett, 2010; Chai et al., 2010; Jimoyiann, 2010). More noteworthy, the initial conceptualization of TPACK has further extrapolated technological content knowledge (TCK), technological pedagogical knowledge (TPK), and technological pedagogical content knowledge (TPCK), which are three potential and additional components that could be closely connected with teaching effectiveness and efficiency. Although the inceptive theoretical framework of TPACK seems promising and attractive, the complex, unclear, and untenable TPACK framework has remained a tricky problem in present studies (Angeli & Valanides, 2009; Archambault & Barnett, 2010; Chai et al., 2010; Jimoyiann, 2010).

In view of these deficiencies, several studies have attempted to further clarify the ambiguous TPACK definition and framework. For example, Schmidt, Baran, Thompson, Mishra, Koehler, and Shin (2009) explored TPACK framework through further investigating pre-service teachers' perception of TPACK. The survey went through factor and item analyses, and expert reviews for content validity. The instrument may help educators design longitudinal studies to assess pre-service teachers' development of TPACK. Seemingly, the study findings of Mishra et al. (2009) are promising and favorable, mainly because they successfully identify a seemingly clear and tenable TPACK structure. Nevertheless, the small sample size of the study greatly hampers the generalizability of the study results.

Chai et al. (2010) addressed 889 pre-service teachers to investigate their perceptions of teachers' TPACK. Although both exploratory factor analysis (EFA) and confirmatory factor analysis (CFA) indicated a robust factor structure with four TPACK components (TK, CK, PK, TPACK), their study findings still failed to probe the other potential and underlying components such as TCK and TPK in the initial TPACK framework. Archambault and Barnett (2010) tested online teachers' perception of TPACK. They revisited and explored the TPACK framework through using a factor analysis based on using a survey with 24 items designed to measure each of the areas described by the TPACK framework. While the framework is helpful from an organizational standpoint, however, it is difficult to separate out each of the domains.

Although contributions to TPACK development are preeminent and undeniable, efforts are still centered on pre-service teachers' perception of their TPACK with few addressing students' perception of their teachers' knowledge. According to Tuan, Chang, Wang, & Treagust (2000), student perceptions of a teacher's knowledge may provide rich information about students' cognition and classroom processes. Jaskyte et al. (2009) suggested both teacher and student views are important sources when we intend to examine teacher knowledge in classroom contexts.

With regard to common students' perceptions of teachers' TPACK, Jang and Chen (2010a) addressed 317 college students to further examine their perceptions of teachers' TPACK. It was found that four factors containing Subject Matter Knowledge (SMK), Instructional Representation and Strategies (IRS), Knowledge of Students' Understanding (KSU), and Technology Integration and Application (TIA) were present in the 33-item TPACK instrument. This four-factor structure somewhat differs from the conceptual TPACK framework proposed by Mishra and Koehler (2006). Jang and Chen (2010a) intended to design a TPACK questionnaire utilizing a transformative approach. That is, use PCK as the core and transform it to further include technology dimensions. Therefore, the PCK domain including SMK, IRS, and KSU factors, and the TIA factor including TK, TCK, TPK, and TPCK components would be connected with teaching effectiveness. Angeli and Valanides (2009) argued the boundaries between some components of TPACK, for example, what they define as TCK and TPK, are fuzzy, indicating a weakness in accurate knowledge categorization or discrimination, and, consequently, a lack of precision in the framework.

METHODOLOGY

This study employs a cross-sectional (pre-test/post-test), mixed-methods design. The cross-sectional design enables us to identify the participant's TPACK profiles, and to track the participant's TPACK development.

Sample and Workshop

The case Physics instructor was recruited from *The 2010 New Faculty TPACK Workshop*, in which a total of eight new college teachers participated as trainees. The teacher was a newly employed full-time physics instructor at the Department of Physics in a university in Taiwan, who mainly taught physics theories-related subjects. With regard to the background of the instructor, John had a two-year postdoctoral research experience in the physics research field and two years of college teaching experience. The instructor was enthusiastic about teaching, gaining new knowledge and willing to make changes. The subject taught by the instructor was "statistical physics" and a total of 70 students were tested. In order to understand the teacher's TPACK development and his actual teaching situation, the questionnaire was distributed to the students at the middle and the end of the semester, respectively.

Instrumentation

This study adopted the 33-item instrument developed by Jang and Chen (in press) to measure university students' perceptions of a teacher's TPACK. A 5-point Likert-type scale response with anchors-, including "1=Never", "2=Seldom", "3=Sometimes", "4=Often", "5=Always", was utilized to measure the variables in each TPACK construct. According to Jang and Chen (2010a), 317 college students participated in the study and their perception of TPACK was further examined. An exploratory factor analysis revealed that four factors containing Subject Matter Knowledge (SMK), Instructional Representation and Strategies (IRS), Knowledge of Students' Understanding (KSU), and Technology Integration and Application (TIA) were present in the 33-item TPACK instrument. SMK refers to students' perceptions associated with the extent to which the teacher demonstrates a comprehension of the subject matter and belief within the discipline. IRS refers to students' perceptions as to the extent to which the teacher uses a representational repertoire including analogies, metaphors, examples, and explanations, and the extent to which the teacher selects teaching strategies. KSU refers to college students' perceptions associated with the extent to which the teacher evaluates student understanding. KSU is used before and during interactive teaching, and at the end of lessons and units. TIA refers to students' perceptions as to the extent to which the teacher selects appropriate technologies and integrates them with content and pedagogy for teaching. TIA includes TK, TCK, TPK, and TPCK, which are potential components that could be closely connected with teaching effectiveness. It was shown that the SMB factor with $\alpha = .943$, IRS factor with $\alpha = .949$, KSU factor with $\alpha = .902$, and TIA factor with $\alpha = .942$ all indicated satisfactory internal consistency.

3.3. Data Collection and Analysis

In order to understand the instructor's TPACK development and the actual teaching situation, the questionnaire was distributed to the students in the middle and at the end of the semester, respectively. In terms of the quantitative data, SPSS statistical software was used to perform a basic descriptive statistical analysis on the questionnaires returned in the middle of the semester, and the teacher was informed of the analytical results, including the mean and standard deviation of each question, the average score of each aspect, and the feedback and suggestions from the students.

The survey employed in this research also included one open-ended question for students to comment on the course. The interviews were employed at the end of the semester to explain and support the statistical analysis given to the instructor by the students. This study used a mixed method design, incorporating both quantitative and qualitative techniques. The quantitative data was statistically analyzed using a paired sample t-test to examine the instructor's TPACK development. The qualitative data was mainly collected from the instructor's interviews and students' opinions of the survey. The qualitative data was a combination of documentary interpretation (Erickson, 1986) and qualitative analysis (Strauss, 1987). The study adopted constant comparison and the triangulation method related to the instructor's interviews, students' opinions and quantitative data to explain the instructor's TPACK change.

FINDINGS

Table 1 shows the descriptive statistics of the students' responses. The statistics of the teacher's TPACK performance in the four categories about the mean scores and standard deviations of both the pre-test and post-tests. All post-test mean scores were higher than that of their pre-test scores. We found that the case Physics teacher's TIA score increased significantly (t = 2.215) from the middle to the end of the semester; however, no significant changes in SMK, IRS, and KSU scores were detected.

Table 1. The statistics of John's TPACK performance on four categories

Category	pre-test		post-test		t
	M	SD	M	SD	
Subject Matter Knowledge	4.42	0.73	4.48	0.58	.400
Instructional Representation & Strategies	3.94	0.78	4.07	0.66	1.123
Knowledge of Students' Understanding	4.09	0.76	4.31	0.61	1.422
Technology Integration & Application	4.00	0.85	4.36	0.74	2.215*

* p<.05.

Subject Matter Knowledge

Besides obtaining the average scores of the four main aspects in the pre-test and post-test, the items in the pre-test and post-test were further compared to understand the teacher's TPACK development under the influence of the four aspects. Table 2 shows the teacher's average score for each question in the aspect of "SMK." According to the t-test analysis results, there is no significant difference between each question in the SMK aspect.

Table 2. The statistics of teacher's TPACK performance on SMK

Item	pre-test		post-test		t
	M	SD	M	SD	
01	4.65	.56	4.58	.58	-.625
02	4.65	.56	4.54	.65	-.827
03	4.46	.76	4.54	.65	.570
04	4.38	.90	4.54	.65	.778
05	4.46	.58	4.54	.65	.493
06	4.04	1.08	4.27	.72	1.237
07	4.50	.95	4.46	.65	-.214
08	4.27	.96	4.42	.64	.811
09	4.46	.95	4.38	.75	-.386
10	4.35	1.02	4.50	.71	1.000

In the SMK aspect, item 01 ("My teacher knows the content he/she is teaching") was found to have the highest score, which exceeded 4.5 on average, in both pre- and post-test; while item 06 ("My teacher explains the impact of the subject matter on society") had the lowest score. It could be inferred that, in this aspect, the teacher had knowledge of the content he taught. On the other hand, the teacher could provide more detailed explanations on the impact of the subject matter on society, as well as its actual applications. Such an approach would be meaningful to the students' learning and the design of teaching contents.

As for relevant qualitative data, we collected students' many opinions from open-ended questions, and most students expressed that they found the teacher able to clearly explain the course contents and answer their questions. For example:

S01: I have forgotten the mathematics taught in high school; however, the teacher makes the subject simple by explaining the contents clearly, and my problems are readily solved.

S05: The teacher is enthusiastic in giving explanations to make the content comprehensible to everyone.

S12: The teacher is always patient in answering our questions.

However, there was one student who indicated in the pre-test that the teacher "sometimes cannot understand the students' questions." John also mentioned this response in the individual interview and provided suggestion methods for solving the problem. For example:

R: One of the students indicated in the pre-test that the teacher "sometimes cannot understand the students' questions." Is it true that you sometimes cannot understand the questions the students asked?

John: When I saw this comment last time, I was quite confused.... I thought I had answered all their questions. Perhaps I didn't give him the answer he wanted. Another reason may be that many mathematical concepts were taught in the first half of the semester, and the student might not know exactly how to ask the question, and thus there was problem in our communication. If a student tells me that he/she cannot understand the content in class, I will explain clearly to him/her after class in my private hours.

[Interview]

As seen above, the student probably did not know how to phrase the question, or had difficulties in communicating with the teacher. However, the teacher was ready to answer the student's question or solve the problem in his private hours. The feedback and opinions from students would also enable the teacher to understand the learning situation and problems faced by the students. In addition, during the interview, the teacher clearly described the overall framework of the course contents. Because some contents were more abstract, the teacher considered them much harder than those of the previous course, thermal physics. For example:

R: This course is statistical physics. What units are included in this semester?

John: The units from the middle to the end of the semester include Fermi gas and Bose gas. These two units actually involve difficult statistics because they are more abstract. For example, Bose Gas is about blackbody radiation, whereas Fermi Gas is about solid-state physics and conduction in metals. I think the contents are more abstract than those of the last semester because this course is a follow-up course of thermal physics in the last semester...Statistical methods are involved in this semester, which makes the contents much harder.

[Interview]

In summary, in this aspect, the teacher expressed his knowledge of the course content, and the students could also perceive the enthusiastic teaching attitude and abundant subject knowledge possessed by the teacher. The teacher could understand his limitations in introducing the subject knowledge, which may confuse the students. As a result, the teacher attached more importance to learning the feedback from students. The teacher not only encouraged students to ask questions in class but also used his private hours to answer their questions. Because the course contents are much harder and more abstract, the teacher used different teaching strategies for explanation, and modified the course contents, taking into consideration students' difficulties in learning, in order to increase their understanding of statistical physics.

Instructional Representation & Strategies

Table 3 shows the average IRS scores of the teacher for each question. The paired sample t-test results indicated no significant difference between each question in the aspect of IRS. In the IRS aspect, item 18 ("My teacher provides an appropriate interaction or good atmosphere") had the highest score, which was 4.46-4.54 on average, in both pre- and post-test, suggesting that the students could perceive the teacher's efforts in creating a suitable teaching or interactive atmosphere. On the contrary, item 15 ("My teacher uses demonstrations to help explain the main concept") and item 17 ("My teacher adopts group discussion or cooperative learning") had lower scores, suggesting that the teacher seldom or never adopted group discussion or cooperative learning as their instructional representation and strategies. An interview was further conducted to understand the teacher's use of teaching strategies.

Table 3. The statistics of teacher's TPACK performance on IRS

Item	pre-test		post-test		t
	M	SD	M	SD	
11	4.23	.82	4.35	.75	.550
12	4.35	.80	4.42	.70	.386
13	4.15	.92	4.42	.64	1.428
14	4.15	1.22	4.15	1.01	.000
15	3.23	1.37	3.69	1.26	1.850
16	4.12	.95	4.15	.88	.189
17	2.31	1.12	2.50	1.21	.723
18	4.54	.761	4.46	.71	-.493
19	4.35	1.02	4.38	.801	.189
20	4.00	1.10	4.19	.90	1.095

As for qualitative data, the students commented that the teacher's teaching method was great, and explicit explanations about abstract concepts were given. On the other hand, John also gave actual examples to explain his instructional representation and strategies, as well as the method adopted, which can promote students' thinking of abstract concepts.

R: How did you make the students understand the professional subject knowledge? Can you take a unit as an example or give an example to explain your major teaching method?

John: I gave an example and drew it on the blackboard. Imagine a spring coil connected to a weight, once it collides with a gas molecule of large energy, the gas molecule will pass through it, but a slow-moving molecule cannot. In the end, the speed of the molecule at one side of the spring coil is getting faster, whereas that of the other side is getting slower. The molecules inside are quite different. Perhaps, this explanation helps students to understand better the concept. [Interview]

According to the above interview record, the teacher mainly used examples or metaphors to explain more abstract concepts, and facilitated the students' thinking. When explaining the concepts, the examples chosen have to tally with the students' life experiences, and even illustrations should be used for explanation so as to enable the students to better imagine or understand the meaning. In addition, the teacher also used some mnemonic skills to help students learn the abstract statistical physics formulas. For example:

R: Students particularly indicated that many formulas are included in this course and it is hard to memorize them. How did you solve this problem?

John: It is true that it is hard to memorize these formulas. I usually teach them with some pithy formulas. For example, I used a block chart for thermal physics in the last semester, which includes the descriptions such as Very Stupid Physics Teacher, and You Have Go Fast. Students can memorize the formulas by memorizing the capital letters. As for this semester, some vocabulary and mnemonic skills were provided to help them.

[Interview]

As for the use of demonstrations, group activities or group discussion, it was also found that the teacher used them less frequently because relevant concepts were more abstract and involved a higher level of knowledge. Hence, students may not be able to generalize and arrange the concepts through discussions, thus accounting for the lower scores of items 15 and 17 for the teacher. For example:

John: It is not easy to apply group discussion to this course. It takes a lot of time simply to explain a very small topic, and it's hard to imagine how much time a group may need for discussion. Because they do not have sufficient prerequisite knowledge, and these concepts, which involve a higher level of knowledge and are much different from those learned in general physics, it is hard for them to obtain any result through group discussion...In addition, these concepts are abstract and usually involve more than two aspects of knowledge; thus, it is hard to conduct a group discussion. [Interview]

In summary, in this aspect, the teacher's use of diversified IRS could help explain the abstract concepts, and assist students to memorize the more abstract formulas and concepts. Although the scores for methods, such as group discussion or cooperative learning, were lower, as a whole, they did not influence students' perception of the changes and efforts made by the teacher in the IRS aspect. Besides using actual examples or metaphors to give explanations, the teacher also used some teaching videos or multimedia animations for discussion on some contents.

Knowledge of Students' Understanding

The teacher's average scores for each item in the aspect of "Knowledge of Students' Understanding" are shown in Table 4. As can be seen, there was no significant difference between each item in the aspect of KSU. However, the average score of each item slightly increased. In addition, it was found that item 26 ("My teacher's assignments facilitate my understanding of the subject") had the highest score in the pre-test, while item 23 ("My teacher's questions evaluate my understanding of the topic") had the highest score in the post-test. Item 25 ("My teacher uses different approaches (e.g. questions and discussion) to find out whether I understand or not") had the lowest score in both the pre- and post-test.

Feedback from the students indicated that they perceived the course content to be moderately difficult, while some found it very difficult due to their lack of relevant background knowledge, and thus, felt more stress in learning. For this problem, the teacher indicated that he had tried to understand the students' difficulties in learning, and provided extra remedial instruction. However, owing to course progress and limited time, it was hard to answer all the students' questions in class. For example:

S15: I have forgotten the mathematics taught in high school; however, the teacher makes the subject simple by explaining the contents clearly, and my problems are readily solved

S23: The teacher is always patient in answering our questions; therefore, we do not have difficulties in learning.

John: When the students told me that they could not understand the contents in class, I would explain to them after class and asked them which part they could not understand. To my knowledge, their mathematical background knowledge was insufficient. For example, it seemed that they did not quite understand concepts including probability, circumference of a circle, permutation and combination, etc...Under such circumstances, it is difficult and time-consuming to explain the contents. I could only ask them to discuss this with me in private.

[Interview]

On the other hand, the teacher also explicitly explained the course evaluation method in the interview. In addition to the traditional written examinations, some extra assignments were arranged for assessment purposes so as to enable the students to practice, as well as to understand, their difficulties in learning. Therefore, the teacher adopted more diversified methods for evaluation, and provided the students with additional opportunities to earn bonus points. In general, the students approved the teacher's evaluation methods.

Table 4. The statistics of teacher's TPACK performance on KSU

Item	pre-test		post-test		t
	M	SD	M	SD	
21	4.23	.86	4.42	.70	.926
22	4.04	1.04	4.35	.63	1.443
23	4.12	.86	4.46	.65	1.735
24	4.12	.77	4.35	.69	1.443
25	3.77	1.03	3.92	.93	.582
26	4.27	.92	4.42	.64	.891

R: Besides the interim and final written examinations, what other evaluation methods are included in this course?

John: There are also quizzes, assignments, and a class presentation. Three quizzes are arranged at present, and assignments are given after the instruction of every chapter. The students are asked to give a presentation of their assignment in class. When they give a presentation, I will rate their performance. The scores of assignments account for 15% of the total score, attendance rate accounts for 5%, and the scores of quizzes account for 5%. I am still adjusting the ratio. I have told them that these might be extra scores, and the scores of interim and final examinations may account for 40%, respectively. In general, the students can all obtain higher grades.

[Interview]

In summary, the teacher could use diversified evaluation methods, understand the students' insufficient prerequisite knowledge and learning difficulties, and improve the teaching contents and methods. The teacher also arranged extra remedial instruction or gave explanations to the students with insufficient prerequisite knowledge and learning difficulties.

Technology Integration & Application

The teacher's average scores for each TIA item are shown in Table 5. As can be seen, the TIA subscale score of the case Physics teacher significantly increased from the pre-test to the post-test, meaning that students perceived that the case Physics teacher increased the incorporation of technology into his teaching. When we examined individual survey items, we found that the scores of items 29 and 32 had the most significant increase. More specifically, item 29 is related to the use of technology for the content or a specific course unit, while item 32 is associated with the amalgam of technology, pedagogy, and content knowledge. The above results indicated that the case Physics teacher not only has knowledge about technology but also knows how to integrate it with course content and pedagogy.

The result indicated that, in the aspect of TIA, there were significant differences in a total of two questions (item 29— "My teacher is able to choose appropriate multimedia and web technologies that enhance his/her teaching for a specific course unit;" item 32— "My teacher is able to integrate content, technology, and teaching methods in his/her teaching."), suggesting that the students generally perceived that the teacher could integrate subject content and teaching methods with technology, and use them properly. As a whole, there was an increase in average scores of all the questions in this aspect.

Table 5. The statistics of John's TPACK performance on TIA

Item	pre-test		post-test		t
	M	SD	M	SD	
27	4.66	.56	4.73	.78	.402
28	3.50	1.17	3.96	1.08	2.004
29	3.62	1.17	4.31	.93	3.048*
30	4.04	1.08	4.35	.94	1.775
31	4.08	.84	4.27	.92	1.154
32	4.04	1.04	4.46	.71	2.186*
33	4.12	.99	4.42	.81	1.443

* $p<.05$.

The case Physics teacher's TIA score on item 27 ("My teacher knows how to use multimedia (e.g. PowerPoint and animation) for teaching.") was the highest (*M* = 4.66 for the pre-test and 4.73 for the post-test). Interestingly, this item also has the largest p-value (.419), meaning that the change from the pre-test to the post-test was the most insignificant. It is possible that the original score was already very high (most students chose "4" or "5"), leaving little room for the score to increase. On the other hand, the case Physics teacher's TIA score on item 28 ("My teacher knows how to use web technologies (e.g. teaching website, Blog, and distance learning) for teaching.") was the lowest (*M* = 3.50 for the pre-test and 3.96 for the post-test), and the score remained statistically the same from the middle to the end of semester.

When we reported this low score to the case teacher during the interview, he explained that although he did not incorporate Web technologies into course activities, he did spend a lot of time online to search for materials and multimedia for teaching. The interview response suggests that there are indeed certain discrepancies between teacher and students' perceptions, especially when some preparation work for teaching is behind the scene. From the researchers' point of view, although the incorporation of Web technologies is a good indicator of technology integration and application (TIA), item 28 does not constitute a necessary condition for effective teaching. More important is the teacher's the ability to integrate technology, pedagogy, and the course content (items 32 and 33).

It was found in the teacher's interview that, after the middle of the semester, the teacher generally increased the use of technological resources, and was able to combine relevant videos with online-Flash animation to add supplementary descriptions to the course, enabling the students to understand the application of relevant abstract concepts through hands-on operating or simulating processes. For example:

R: As opposed to internet technology, you obtained higher scores for the use of multimedia. Why is that? Besides using videos, do you use other teaching technologies or tools?

John: In general, I use videos more frequently, and there is a great difference between the pre-test and post-test. One of the reasons is that before the middle of the semester, the focus of the course is to talk about more theories, which are more difficult...However, after the middle of the semester, the focus of the course is to apply the theories; and thus, some actual phenomena can be used as examples. The course contents of the first half of this semester are theories, whereas those of the second half of this semester are theories and their actual applications, which may explain the difference in my use of technology. Moreover, videos are more likely to be used in the second half of the semester.

R: As for the performance of technology, there are significant differences in some questions and the average scores of all the questions increased...Would you give a more explicit explanation about the technologies used, such as software or animation to strengthen the description about technology integration and application.

John: In addition to videos, I also looked for some flashes as teaching materials. For example, when I give lectures on statistical distribution, such as Maxwell, Bose distribution curves...blackbody radiation. When the temperature of light and CRE are used as the coordinates, the temperature parameter can be adjusted. For example, the illumination of motorcycle headlamp is 7100K, that of the general sunlight is 6500K, and that of yellow light is 3500-3600K. If Flash is used to change the coordinates of temperature, the distribution curve will change.

[Interview]

It was also verified that the teacher was able to choose different teaching technologies for application according to the differences in units or contents. For example, when explaining abstract concepts, the teacher might use PPT to summarize the concepts and give explanations.

However, when explaining an actual application or relevant experiments, the teacher might increase the use of relevant videos to help students understand them. Moreover, this study further probed into the teacher's resources of videos and teaching technologies. For example:

R: Did you find the teaching videos at school?

John: There are many such videos on Youtube and I found most of them directly on the Internet.

R: Will you edit or arrange the flashes and videos found online, or use them directly?

John: After downloading these videos, I would edit them slightly, convert the files, and remove irrelevant parts. And then, I would link them to the PPT; therefore, the students sometimes may obtain the PPT with hyperlinked videos. [Interview]

In summary, in this aspect, the teacher clearly exhibited that he could make use of multimedia or Internet technologies to teach students or collect teaching materials. Moreover, the teacher could choose appropriate technological teaching tools, taking into consideration the course units, contents, and students' learning methods.

Although the students perceived that the teacher used internet technologies (e.g. websites, blogs) less frequently, the teacher actually obtained relevant educational information on the Internet and integrated it with his teaching materials. Such a teaching method had concretely been exhibited the teacher's application of web-based instruction; however, the students could not perceive this fact in class.

Discussion and Implications

The major contribution of this study used an instrument for assessing university students' perceptions of a science teacher's TPACK. The survey was used to understand the overall teaching performance of the teacher and provide him with teaching reflection. Compared with a traditional semester-end evaluation, which could only collect rather few opinions (Jang, 2011), the design of this study could also collect students' many opinions from open-ended questions and provide a diagnostic function to allow the new teacher to find out his changes or differences after a period of teaching, and make reflective thinking as well as timely modifications.

This study found that students' perceptions of the case physics teacher's TPACK had enhanced through workshops and reflective feedback. University teachers are the authority in the traditional classroom and can easily have self-centered thinking. Using the survey, the researchers can determine whether the teacher achieves the expected goal by understanding students' perceptions (Jang, 2011; Tuan, Chang, Wang & Treagust, 2000). Moreover, the teacher can gain more teaching and learning experience by participating in the investigation. The teacher's TPACK development was evaluated according to judgments of all the students and was not restricted to a few observations or interview data (De Jong, Van Driel, & Verloop, 2005). However, the survey has some limitations. The quantitative presentation does not investigate the factors behind the teacher's development in depth. In this study, in addition to the scale, the researchers collected students' qualitative data in various ways, such as interviews, workshop records, and the open-ended opinions of students, to reveal whether the teacher's TPACK had improved.

In this study, the case Physics teacher's TIA score significantly increased from the middle to the end of semester; however, no significant changes had been detected for the SMK, IRS, and KSU categories. The teacher fully exhibited his TPACK development so that he could make use of multimedia or Internet technologies to teach students during this semester. Magnusson et al. (1999) stressed that the development of PCK is determined by the content to be taught, the context in which the content is taught, and the way the teacher reflects on his teaching experiences. Although the teaching experience was taken as the primary source of TPACK development, reflection is also a critical component of professional development. Reflection therefore emerges as another important element for novice teachers in developing expertise in their practice, and is central to their accepting more responsibility for their actions. Nilsson (2008) emphasized the role of teaching experience and reflection as a way of better understanding the complex entities that constitute a knowledge base for teaching. She draws attention to the value of teachers participating in experiences that might contribute to the development of their PCK. This supports the view of TPACK development as a process of transformation in this study.

On the other hand, the teacher may take more time to develop or change his IRS, KSU and SMK. Nonetheless, there were slight increases in SMK, IRS, and KSU scores in the case teacher, even though they were not statistically significant. Longitudinal studies or cross-sectional studies with a longer time span may better reflect the changes in SMK, IRS, and KSU scores of college teachers; and hence, future studies should be conducted in this direction. We also suspect that the insignificant results in the SMK, IRS, and KSU dimensions are partially due to the nature of the survey instruments. Quantitative survey items tend to be generic in order to encompass all dimensions and constructs for portraying general tendencies. However, PCK and TPACK are such broad and amorphous concepts that they contain many possible dimensions and details (Baxter & Lederman, 1999). Therefore, it is practically difficult to design questionnaire items that capture all the subtlety in teachers' knowledge, belief, and actions in various learning contexts. As such, we recommend that qualitative data should also be collected to supplement our TPACK instrument if future studies intend to trace the development of college teachers' TPACK. Perhaps most important, we plan to conduct classroom observations of college teachers to evaluate the level of TPACK demonstrated in their classrooms and then investigate how scores on the TPACK instrument predict classroom behaviors.

Besides explaining and discussing the four main aspects and each item, several suggestions are proposed in the research process as reference to revisions of future questionnaires according to the case study. First, during the interview process, the teacher specifically indicated that authentic feedback and opinions from students can indeed urge a teacher to reflect and make improvements. In addition, the teacher or researchers should encourage the students to answer the open-ended items in the questionnaire as an explanation of the scores given on the quantitative part. Finally, the authors plan studies designed to further validate and revise the instrument. We plan to periodically administer the survey throughout novice college teachers programs, using the results to inform researchers of specific times or events when each knowledge domain is developed. This information will provide valuable insight into the development of TPACK and provide program feedback on effective approaches in encouraging this development.

REFERENCES

Angeli, C. & Valanides, N. (2009). Epistemological and methodological issues for the conceptualization, development, and assessment of ICT–TPCK: Advances in technological pedagogical content knowledge (TPCK). *Computers & Education, 52*(1), 154-168.

Archambault, L. M. & Barnett, J. H. (2010). Revisiting technological pedagogical content knowledge: Exploring the TPACK framework. *Computers & Education, 55*(4), 1656-1662.

Chai, C. S. Koh, J. H. L., Tsai, C.-C. & Tan, L. L. W. (2011). Modeling primary school pre-service teachers' Technological Pedagogical Content Knowledge (TPACK) for meaningful learning with information and communication technology (ICT). *Computers & Education*, 57, 1184-1193.

Chai, C. S., Koh, J. H. L. & Tsai, C.-C. (2010). Facilitating pre-service teachers' development of Technological, Pedagogical, and Content Knowledge (TPACK). Educational Technology & Society, 13 (4), 63–73.

De Jong, O., Van Driel, J.H., & Verloop, N. (2005). Preservice teachers' pedagogical content knowledge of using particle models in teaching chemistry. *Journal of Research in Science Teaching, 42*(8), 947–964.

Jang, S. J. & Chen, K. C. (2010a, December). Developing an instrument for assessing TPACK of college science teachers through student perception. *2010 Global Chinese Conference on Science Education, Institute of Education.* CN: Hong Kong.

Jang, S. J. & Chen, K. C. (2010b). From PCK to TPACK: Developing a transformative model for pre-service science teachers. *Journal of Science Education and Technology, 19*(6), 553-564.

Jang, S. J. (2010). Integrating the interactive whiteboard and peer coaching to develop the TPACK of secondary science teachers. *Computers & Education, 55*(4), 1744-1751.

Jang, S. J. (2011). Assessing college students' perceptions of a case teacher's pedagogical content knowledge using a newly developed instrument. *Higher Education, 61*(6), 663-678.

Jaskyte, K., Taylor, H. & Smariga, R. (2009). Student and faculty perceptions of innovative teaching. *Creativity Research Journal, 21*(1), 111-116.

Jimoyiann, A. (2010). Designing and implementing an integrated technological pedagogical science knowledge framework for science teachers' professional development. *Computers & Education, 55*(3), 1259-1269.

Kopcha, T. J. (2010). A systems-based approach to technology integration using mentoring and communities of practice. *Educational Technology Research & Development, 58* (2), 175-190.

Kramarski, B. & Michalsky, T. (2010). Preparing preservice teachers for self-regulated learning in the context of technological pedagogical content knowledge. *Learning and Instruction, 20*(5), 434-447.

Magnusson, S., Krajcik, J. & Borko, H. (1999). Nature, sources, and development of pedagogical content knowledge for science teaching. In J. Gess-Newsome & N. Lederman (Eds.), *Examining pedagogical content knowledge* (pp. 95-132). Netherlands: Kluwer.

Mishra, P. & Koehler, M. J. (2006). Technological pedagogical content knowledge: A framework for teacher knowledge. *Teachers College Record, 108*(6), 1017-1054.

Niess, M. L. (2005). Preparing teachers to teach science and mathematics with technology: Developing a technology pedagogical content knowledge. *Teaching and Teacher Education, 21*(5), 509-523.

Nilsson, P. (2008). Teaching for understanding: The complex nature of pedagogical content knowledge in pre-service education. *International Journal of Science Education, 30*(10), 1281-1299.

Schmidt, D. A., Baran, E., Thompson, A. D., Mishra, P., Koehler, M. J. & Shin, T. S. (2009). Technological Pedagogical Content Knowledge (TPCK): The development and validation of an assessment instrument for preservice teachers. *Journal of Research on Technology in Education, 42*(2), 123-149.

Shulman, L. S. (1986). Those who understand: knowledge growth in teaching. *Educational Researcher, 15*, 4–14.

Shulman, L. S. (1987). Knowledge and teaching: foundations of the new reform. *Harvard Educational Review, 57*, 1–22.

Tuan, H., Chang, H., Wang, K., & Treagust, D. (2000). The development of an instrument for assessing students' perceptions of teachers' knowledge. *International Journal of Science Education, 22*(4), 385-398.

Appendix A: Students' Perceptions of University Teachers' TPACK

Directions for students:

This questionnaire contains five statements about teaching practices which could take place in this class. You will be asked how often each practice takes place. There are no "right" or "wrong" answers. Your opinion is what is wanted. Think about how well each statement describes what this class is like for you. You will be asked to describe freely your personal comments to the course in the end. Be sure to give an answer for all questions.

Draw a circle around

1. if teaching practice takes place Never
2. if teaching practice takes place Seldom
3. if teaching practice takes place Sometimes
4. if teaching practice takes place Often
5. if teaching practice takes place Always

A. Subject Matter Knowledge (SMK)

1. My teacher knows the content he/she is teaching.
2. My teacher explains clearly the content of the subject.
3. My teacher knows how theories or principles of the subject have been developed.
4. My teacher selects the appropriate content for students.
5. My teacher knows the answers to questions that we ask about the subject.
6. My teacher explains the impact of subject matter on society.
7. My teacher knows the whole structure and direction of this subject matter.
8. My teacher makes me clearly understand objectives of this course.
9. My teacher pays attention to students' reaction during class and adjusts his/her teaching attitude.
10. My teacher's belief or value in teaching is active and aggressive.

B. Instructional Representation & Strategies (IRS)

11. My teacher uses appropriate examples to explain concepts related to the subject matter.
12. My teacher uses familiar analogies to explain concepts of the subject matter.
13. My teacher's teaching methods keep me interested in this subject.
14. My teacher provides opportunities for me to express my views during class.
15. My teacher uses demonstrations to help explain the main concept.
16. My teacher uses a variety of teaching approaches to transform subject matter into comprehensible knowledge.
17. My teacher adopts group discussion or cooperative learning.
18. My teacher provides an appropriate interaction or good atmosphere.
19. My teacher creates a classroom circumstance to promote my interest for learning.
20. My teacher prepares some additional teaching materials.

C. Knowledge of Students' Understanding (KSU)

21. My teacher realizes students' prior knowledge before class.
22. My teacher knows students' learning difficulties of the subject before class.
23. My teacher's questions evaluate my understanding of a topic.
24. My teacher's assessment methods evaluate my understanding of the subject.
25. My teacher uses different approaches (questions, discussion, etc.) to find out whether I understand.
26. My teacher's assignments facilitate my understanding of the subject.

D. Technology Integration & Application (TIA)

27. My teacher knows how to use multimedia (e.g. PowerPoint and animation, etc.) for teaching.
28. My teacher knows how to use web technologies (e.g. teaching website, Blog, and distance learning) for teaching.
29. My teacher is able to choose multimedia and web technologies which enhance his/her teaching for a specific course unit.
30. My teacher is able to use technology to enhance our understanding and learning of lessons.
31. My teacher is able to use technology to enrich the teaching content and materials.
32. My teacher is able to integrate content, technology, and teaching methods in his/her teaching.
33. My teacher is able to choose diverse technologies and teaching methods for a specific course unit.

Chapter 6

USING INTERACTIVE WHITEBOARD TECHNOLOGY AND OBSERVATION TO DEVELOP THE TPACK OF SCIENCE TEACHERS

ABSTRACT

Interactive Whiteboard (IWB) technology and the Technological Pedagogical and Content Knowledge (TPACK) of teachers are important issues. However, it would appear that the means for training science teachers to use the IWB and assessing the TPACK of science teachers have yet to be fully investigated. This study provides a theoretical framework for the IWB-TPACK model and develops evaluative criteria from an integrative view to assess the case science teacher's TPACK. The case teacher and her 32 seventh-grade students participated in this study. Two instances of classroom teaching observations were conducted. The data collected from the evaluative observation, reflective journal and interviews were used to evaluate the TPACK development. Results showed that the case teacher's TPACK improved on the whole, especially in technology-related knowledge and appropriate use of animations and videos as part of the IWB. Furthermore, observation produced teaching reflection with the case teacher, who used the IWB to enhance teacher-student interaction and evaluate student knowledge. The research implications of this study are provided along with suggestions.

Keywords: Interactive Whiteboard; TPACK; Evaluative criteria; Teaching observations

1. INTRODUCTION

There has been an increasing tendency to enhance teachers' application ability of educational technology, as exemplified by the "Whitepaper for Educational Technology in K-12 Education." of the Ministry of Education in Taiwan (2008). One of the goals associated with this whitepaper was to increase teaching quality by integrating technology and pedagogy. More specifically, the infrastructures in this whitepaper are centered on encouraging teachers to use educational technology, inspire student's creative thinking ability, share technology resources with students and ensure that students have technology learning opportunities. As a result, each teacher is expected to know how to effectively apply educational technology in their curriculum designs, teaching implementation, teaching management, and teaching evaluation (Jang, 2008). Taiwan also recognized the potentials of Interactive Whiteboards (IWBs) and subsidized schools to purchase these instructional tools. However, there are still many science teachers who have not previously used IWBs (Jang, 2010). Some possible reasons might include difficulties of teaching material design, budget constraints, classroom space limitations, and problems in manipulation and training. Although the IWB is a promising pedagogical tool, several issues emerged during the implementation process. This is not surprising, as the introduction of a new technology is usually not accompanied by an adequate understanding of what its implementation might imply for pedagogy (Hennessy, 2006; Wellington, 2005). This echoes the finding of Smith et al. (2005), that the use of this technology is limited by the lack of adequate training for teachers beyond that which is initially provided by IWB companies and suppliers, and by the difficulty of physically placing the IWB in the classroom in order to maximize viewing by the whole class. Teachers also note that they needed considerably more time to prepare IWB lessons than for regular lessons (Higgins, Beauchamp & Miller, 2007).

A growing number of studies are now available to shed some light on the integration of technology into pedagogy. In some of these studies, researchers have considered the role of IWBs increasing in student motivation (Glover, Miller, Averis & Door, 2007; Hall & Higgins, 2005; Hennessy, Deaney, Ruthven, & Winterbottom, 2007; Schmid, 2008; Slay, Sieborger, & Hodgkinson-Williams, 2008; Torff & Tirotta, 2010), and student performance (Lopez, 2010; Sedig & Liang, 2006). In this view, the IWB is an innovative teaching tool which is designed to replace traditional blackboards and whiteboards, and make up for the lack of teaching materials. Thus, IWBs are expected to enhance both teaching quality and learning efficiency. More specifically, science teachers can directly demonstrate an abstract concept or simulate a situation corresponding to the subject content. As a result, teacher-student interactions are increased in the classroom.

From a more theoretical perspective, development of the Technological Pedagogical and Content Knowledge (TPACK) framework builds on the earlier work of Shulman (1986), where he described Pedagogical Content Knowledge (PCK) as the intersection between content knowledge and pedagogical knowledge. TPACK is the integration of subject matter knowledge development with the development of technology and of knowledge of teaching and learning (Niess, 2005). Mishra and Koehler (2006) describe a framework for teacher knowledge of technology integration, originally called Technological Pedagogical Content Knowledge (TPCK), now known as TPACK. They argue that the central construct of TPACK is not merely a straightforward combination of knowledge about content, pedagogy and technology, but rather an emergent form of transformative knowledge that truly integrates the components into new forms that have the potential to maximize the effectiveness of educational technology in the classroom.

While TPACK is applicable to any educational technology, this study focuses on IWBs and the extent to which in-service secondary teachers are able to develop teaching and learning activities that demonstrate the integration of their content knowledge, pedagogy and IWB technology. Researchers might do well to explore how best to use this powerful teaching tool (Jang, 2010), and how to train teachers to use it effectively (Holmes, 2009). Much research indicates that observation can also have many positive effects on teachers' professional development, such as teaching self-examination ; trying new technology or methods in teaching; and confirming, adjusting or improving their existing teaching methods and contents (Anderson, Barksdale & Hite, 2005; Powell & Napoliello, 2005). Observation for in-service teachers can be seen as a method of exploring the ability of an effectively instructional and technological teacher (Jang, 2008). It has also been found that teachers' own learning experiences with technology are a major determinant as to the extent to which they incorporate technology when teaching (Crison, Lerman & Winbourne, 2007; Jang & Chen, 2010). However, there has not yet been a full investigation of how to train science teachers to use IWBs or an assessment of the TPACK of teachers (Graham et al., 2009; Schmidt et al., 2009). Therefore, in this study, we propose a theoretical framework for an IWB-TPACK model and develop an evaluative tool to assess the secondary science teacher's TPACK. In order to help the case science teacher practically integrate IWBs into his/her teaching, we used peer observation to stimulate and evaluate the development of the case teacher's TPACK, as described in detail in the Methods section.

2. Theoretical Framework

2.1. TPACK Framework

Mishra and Koehler (2006) quoted Sir William Henry Bragg's proposition that "The important thing in science is not so much to obtain new facts as to discover new ways of thinking about them (p.1017). " TPCK was introduced to the educational research field as a theoretical framework for understanding the teacher knowledge required for effective technology integration. However, at the 9th Annual National Technology Leadership Summit, the TPCK framework acronym was changed to TPACK (pronounced "tee-pack") to make it easier to remember and to form a more integrated whole for the three kinds of knowledge addressed: technology, pedagogy, and content (Thompson & Mishra, 2007).

Koehler and Mishra (2005) proposed the TPCK framework, which contains seven elements or categories: 1) technology knowledge (TK), 2) content knowledge (CK), 3) pedagogical knowledge (PK), 4) pedagogical content knowledge (PCK), 5) technological content knowledge (TCK), 6) technological pedagogical knowledge (TPK), and 7) TPCK. According to Mishera and Koehler (2006) and their colleagues, the TPACK framework was built on Shulman's PCK to include technology knowledge as situated within content and pedagogical knowledge in a specific context (Mishera & Koehler, 2009; Schmidt, et al., 2009). This context might include students' prior knowledge and learning difficulties, students' interaction, student knowledge evaluation, etc. (Grossman, 1990; Shulman, 1987; Tamir, 1998). Therefore, the researchers consider that the TPCK's major core is the intersection of C, P and T, which is different from the whole package of TPACK, as shown in Figure 1. The latter emphasizes on integrated knowledge which includes not only P, C, T, PC, TC, TP and TPC, but also more context-based knowledge.

Koehler, Mishra, and Yahya (2007) used an approach based on discourse analysis to track the development of TPACK. Analyzing the conversations of teachers working in design teams, they have tracked the development of each of the seven components of TPACK over the course of a semester. This approach, however, is especially time-consuming and is methodologically specific to the unique context in which it was used (i.e., semester-long design experiences). Angeli and Valanides (2009) explored the use of design-based performance assessments embedded into course sequences. They used self-assessment, peer assessment, and expert assessment of these design-based performances as formative and summative assessments of teachers' understanding.

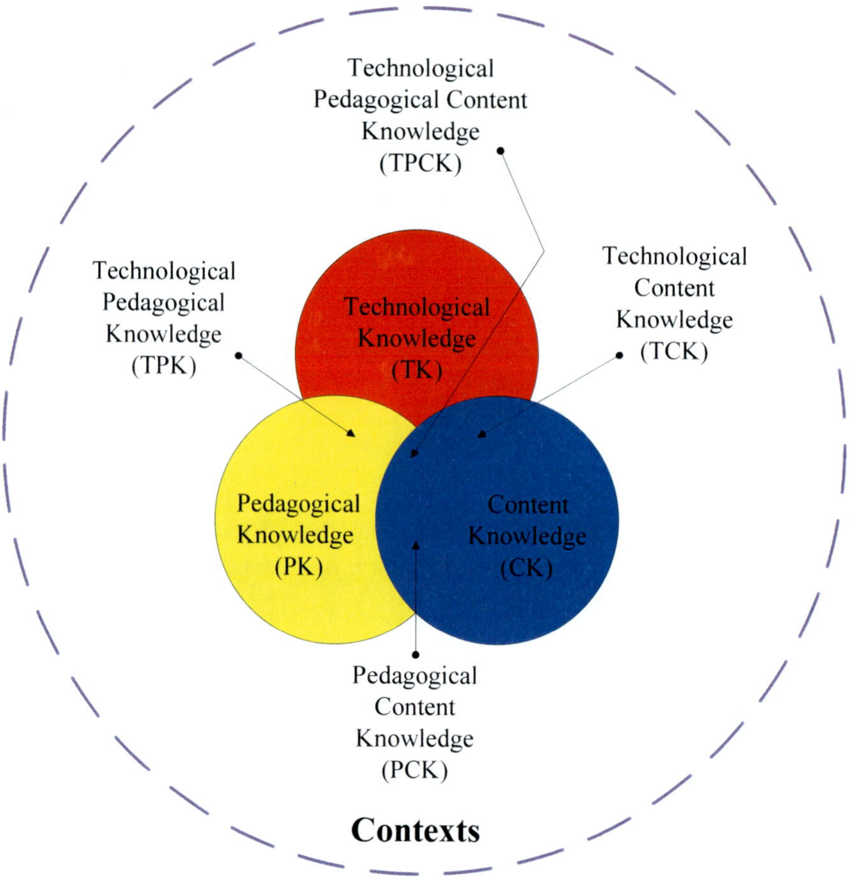

Figure 1. The framework of revised TPACK. (The intersection of C, P and T is TPCK).

Graham, et al. (2009) designed a survey to measure elementary science teachers' TPACK confidence. The survey measures four components of the TPACK framework: 1) technological knowledge (TK), 2) technological pedagogical knowledge (TPK), 3) technological content knowledge (TCK), and 4) TPCK. Thirty-one items were included in the survey, with 11 items assigned for TK, 7 items for TPK, 5 items for TCK, and 8 items for TPCK. The survey items were clearly stated and each targets a specific task. Though the reliability was reported to be high in each of the four domains, the survey had not gone through an item analysis and factor analysis to establish its construct/factorial validity. Moreover, the instrument appears to be technology-oriented because the four measured components (TK, TPK, TCK, and TPCK) are all associated with technology. But surprisingly, content knowledge (CK), pedagogical knowledge (PK), and pedagogical content knowledge (PCK) were ignored in this instrument.

A more rigorously constructed instrument is found in Schmidt et al.'s (2009) study, which intends to assess pre-service teachers' TPACK knowledge structure. That questionnaire contained 47 items spread over the seven TPACK domains: 7 TK items, 12 CK items, 7 PK items, 4 PCK items, 4 TCK items, 5 TPK items, and 8 TPCK items. The survey went through factor and item analyses as well as expert reviews for content validity. Reliability tests revealed acceptable-to-excellent results for the seven subscales. According to the authors, the instrument will help educators design longitudinal studies to assess pre-service teachers' development of TPACK.

2.2. Interactive Whiteboard Technology

Various groups of researchers have worked with the relationship between the use of IWBs and the impact on teaching (Glover, Miller, Averis & Door, 2005, 2007; Higgins, Beauchamp & Miller, 2007; Holmes, 2009; Jang, 2010; Shenton & Pagett, 2007; Smith, Higgins, Wall & Miller, 2005). Smith et al. (2005) have identified several potential benefits for teachers to integrate IWBs into their teaching, including greater flexibility and variability, and the ability to deliver multimedia or multimodal presentations with greater efficiency within the presentation due to the touch sensitive screen.

The integration of interactive electronic whiteboard and multimedia technology (Gillen, Littleton, Twiner, Staarman & Mercer, 2008) uses an exhibition platform of real items for instant digitization, and these can be projected onto the interactive electronic whiteboard, which allows for annotations on the e-teaching materials. The whole instructional process can be automatically saved in an electronic notebook, easily stored for usage and after-class review. This type of new instruction emphasizes "bilateral interaction," in which the proposed questions are used to clarify problems or understand the learning effects of students. The instant feedback system can be used to quickly achieve the learning objective of bilateral interaction. IWBs can lead students to the process of internalization by offering available multimedia, simulations, and modeling (Lopez, 2010).

Many teachers consider the potential of the IWB as more than a teacher resource box (Warwick & Kershner, 2008) that is better able to capture and hold the learners' attention, thereby leading to increased student motivation and the ability to teach at a faster pace (Gillen et al., 2008 ; Shenton & Pagett, 2007). Positive outcomes have been achieved when teachers have been able to incorporate the capabilities of the IWB technology into their existing effective pedagogical approaches. In addition, the use of IWBs supports teachers' planning and development of teaching resources, allows teachers to model appropriate technology skills for the students, and improves interactivity and student participation in lessons (Holmes, 2009).

There are also some studies on the use of IWBs that point to the increases in student engagement. However, the degree to which positive outcomes are achieved heavily depends on individual teachers (Glover et al., 2007). In particular, it has been found that good teachers used a wide variety of resources to keep students on task for most of the lessons by providing clear, visual representations of concepts and activities that incorporated immediate feedback for students. Glover et al. (2005) pointed out that the advantage of interactive electronic whiteboards is in increasing interaction between teachers and students, as well as among students. When applying it to teaching, teachers can use the storage and reappearance of information using functions such as "drag" to make explanations. Students can observe how others interact with the electronic whiteboard, compare their own operational processes and experiences with those of others, and express the learning process, thereby increasing knowledge exchange and interaction among peers.

In summary, instructional strategies that utilize interactive electronic whiteboards are divided into two types. First, there is the teacher-centered exhibition platform: a model that emphasizes the transmission and learning of knowledge. In this learning process, there are fewer opportunities for teacher-student interactive learning. Teachers use the electronic whiteboard system together with teaching materials in the electronic whiteboard database for classroom instruction, taking advantage of its simplicity and efficiency, as well as its pictorial and dynamic character. Second, there is the learning platform with teacher-student interaction: a student-centered process of learning, in which there are more opportunities for interactive learning between the teacher and student or among students. The teacher uses the electronic whiteboard system to develop autonomously or supplement with instructional materials from the electronic whiteboard database to design classroom instruction. Assistance from the electronic whiteboard can accelerate the students' proactive exploration for related topics, and they are able to control the flow and tempo of learning. At this level, the teachers express the characteristics of the electronic whiteboard as simple, quick, and interactive (Glover et al., 2005, 2007).

2.3. Observation of Teaching

Peer observation of teaching is a process that encourages reflection on teaching practice, identifies developmental needs, and fosters debate and dissemination around best practice (Hammersley-Fletcher & Orsmond, 2004). Observation has four important features: (1) stimulate teachers' self-criticism, (2) promote interactive quality in teaching, (3) discover the facts of teaching in the classroom, and (4) criticize constructively the effect of curriculum design (Waxman, 1995). Observation can help teachers to organize the theories they had learned for practical implementation in a class and provide them with an opportunity to contrast theories and practices. Norris (1984) describes the observation skills as consisting of three components: (1) make good preparation for observation, (2) fill out the observation journal to a high standard, and (3) accurately assess the credibility of the observation. These three components of guided observations allow in-service teachers to identify and focus on a single aspect of teaching or learning. When teachers view classrooms through this lens, it may provide them opportunities to draw conclusions that result in improved teaching practices (Anderson et al., 2005; Powell & Napoliello, 2005). In addition, these three components force the observers to introspect and explain their observation to examine their own standpoint and prejudice.

2.4. Developing the IWB-TPACK Model

In the present study, TPACK was investigated within a particular type of technology-based learning environment, the IWB, which is conducive to developing teaching and learning knowledge. Higgins et al. (2007) found that IWBs are ideally suited to whole class presentations, providing a variety of representations, and IWBs have the potential to cater to a variety of learning styles, while also addressing some classroom management issues often prevalent when technology is used. These observations align with the TPACK notion of a transformative learning experience and integrative teacher knowledge. Science teachers with TPACK know how to incorporate the three different knowledge domains of Technology, Pedagogy and Science into different situations and themes. Therefore, science teachers with teaching knowledge are expected to have TPACK as an integrated knowledge (Jimoyiannis, 2010). They are further expected to know how to design multidimensional technology-based curriculums and appropriately develop teaching strategies and representations for providing students with comprehensive knowledge.

Jang (2010) found that science teachers used IWBs as instructional tools to share their subject matter knowledge and to express students' understanding. In addition, the IWBs helped the science teachers who encountered teaching difficulties in the traditional classroom to better implement their representational repertoires and instructional strategies. Finally, the proposed model of integrating IWBs and peer coaching could develop the TPACK of science teachers. Holmes (2009) examined the lesson activities developed by a group of final year undergraduate secondary pre-service teachers. The analysis was guided by the TPACK framework of Mishra and Koehler (2006). The study revealed that the pre-service teachers were able to effectively plan to integrate IWB features within their mathematical lessons and, consequently, demonstrated developing TPACK. They found that the primary benefits of the technology related to its potential to engage students with varied visual representations and virtual manipulation that can help conceptual understanding.

The Interactive Whiteboard-Technological Pedagogical and Content Knowledge (IWB-TPACK) theoretical model of this study based on the Mishra and Koehler (2006) includes eight different theoretical dimensions as shown in Figure 1. Content knowledge (CK) includes science facts, concepts and principles of the subject. Pedagogical Knowledge (PK) describes the collected practices, strategies, procedures, and methods of teaching and learning. Technology Knowledge (TK) encompasses modern technologies such as computers, the internet, digital video, and IWB. This study considers how to integrate the IWB into teaching activities. Pedagogical Content Knowledge (PCK) refers to the knowledge of pedagogy that is applicable to the teaching of specific content. Technological Pedagogical Knowledge (TPK) refers to the knowledge of how the IWB technology can support pedagogical goals. Technological Content Knowledge (TCK) refers to the knowledge of how the IWB technology can create new representations for specific content. Technological Pedagogical Content Knowledge (TPCK) refers to the knowledge of how to integrate the teaching content, IWB technology and teaching methods for a specific course unit. Context Knowledge (CxK) refers to knowledge needed to pay attention to students' learning difficulties of subject, and use the IWB to evaluate students' understanding.

Angeli and Valanides (2009) argued that the boundaries between some components of TPACK, for example, what they define as TCK and TPK, are fuzzy, indicating a weakness in accurate knowledge categorization or discrimination, and, consequently, a lack of precision in the framework. Archambault and Barnett (2010) revisited and explored the TPACK framework through the use of a factor analysis based on using a survey with 24 items designed to measure each of the areas described by the TPACK framework. While the framework is helpful from an organizational standpoint, it is difficult to separate out each of the domains, however. Three major factors become evident, and the only clear domain that distinguishes itself is that of technology. It seems that the TPACK framework experiences the same difficulty as Shulman's conception of pedagogical content knowledge. Thus, this study used integrative views associated with the development of TPACK with the IWB technology and the knowledge of teaching and learning. Accordingly, this current study focuses on IWB technology to develop the TPACK of in-service secondary teachers.

3. Research Method

3.1. Research Design

This study deployed an evaluative tool to explore the TPACK literacy of IWB technology in secondary school teachers. A literature review was first performed to construct the contents of the tool, and a classroom observation was carried out for two instances in which the case teacher taught science units. Data collected from the evaluative observation, reflective journal and interviews were used to evaluate the TPACK development for the case teacher.

3.2. Samples

The case teacher Amy and her 32 seventh-grade students participated in this study. Amy has a degree in Biology and has five years of teaching experience. She pays much attention to students' reactions to her teaching and therefore wishes to learn new teaching methods and how to use IWB technology. The academic performance of the students showed a normal distribution. The teaching materials were based on the one-outline-multiple-textbook policy of the Grades 1-9 Integrative Curriculum. At this age, students have fairly good desire for expression and asking questions, but the insufficiency (or wide disparity) in their knowledge backgrounds limits their thinking, so their life experience should be broadened.

3.3. Context

In this study, two classroom observations were conducted. The case teacher instructed her class, and one expert (the researcher) and three peer science teachers carried out the observation. These three science teachers, with over ten years teaching experience, served as professional mentors in the natural science domain . The topic unit of the first classroom observation was "the digestive system" which took about 45 minutes. The questionnaire scale developed for this study was used for evaluation, augmented with open and concrete personal opinions. It was then followed by an evaluation meeting of about 90 minutes. In this meeting, the instructor explained the features of this particular instance of instruction, as well as the preparation process, and the science teacher peers offered feedback on their observations by providing constructive suggestions. The instructor was interviewed right after the evaluation meeting. Six weeks later, the case teacher conducted the second instance of classroom instruction, with the unit theme of "endocrine system" which also took about 45 minutes. It was similarly observed and rated by one expert and three teacher peers, who also gave open and concrete evaluations. The instructor was also interviewed again right after the second instance of classroom instruction.

3. 4. Data Collection and Analysis

The data collected consisted of: (1) the first and second observation. Two forms of assessment occurred to monitor the TPACK development of the case teacher, namely, expert assessment and peer assessment; (2) the reflective journal written by the science teacher throughout the overall process of the study, and about the comments and suggestions that took place during the evaluative meeting; and (3) the interviews which served to gain a deeper understanding of the science teacher's conceptions.

The Evaluative Criteria for the science teacher's TPACK using the IWB technology was developed based on the IWB-TPACK model of this study. We constructed the important evaluative criteria as the integrative view (package), including one item for each dimension; therefore, there was evaluative criteria for eight dimensions with a total of eight items as in shown table 1. In order to ensure its readability and comprehensibility, the draft questions have been reviewed by science teachers to correct any ambiguity in narration. Then two experts in science education and one scholar in teaching technology checked and verified the form. After retrieving evaluative criteria, reliability and validity tests were conducted according to the rules of the questionnaire compilation. Reliability in this evaluation form was based on "reliability of internal consistency" and "expert validity." The evaluative criteria were based on a statement whether or not the observers agree with the teaching science teacher's TPACK performance during the class. If they chose to "agree" with the statement, they needed write the comments to support their evaluations.

Table 1. Evaluative criteria related to TPACK

Dimensions	Items
CK	1. The teacher can clearly explain the subject content.
PK	2. The teacher can use appropriate representations (e.g. pictures and models) to explain scientific concepts.
TK	3. The teacher knows how to integrate the interactive whiteboard into teaching activities
PCK	4. The teacher's teaching methods keep students interested in the content.
TPK	5. The teacher knows how to use the interactive whiteboard to explain and present the teaching contents.
TCK	6. The teacher knows how to use the interactive whiteboard to enrich the teaching content and materials.
TPCK	7. The teacher knows how to integrate the teaching content, the interactive whiteboard and teaching methods for a specific course unit.
CxK	8. The teacher can use the interactive whiteboard to evaluate students' understanding.

The interview was conducted to explore the extent of transformation associated with the science teacher's TPACK. According to the information gathered from the interviews, the researchers wished to discern the possible discrepancies of the case teacher's view and peer observations. For data analysis, this study adopted the procedures suggested by Patton (1990), which are: 1) collect the original data; 2) organize, categorize, and edit the original data, so that the data files are easily distinguished and accessible; and 3) integrate and summarize important categories from the research question and types of findings so they can be analyzed in an in-depth analysis. The data analysis is a continuous interpretation and decoding of these parts (Erickson, 1986). After comparing such data and after executing triangulation, the researchers can interpret the case teacher's professional growth in terms of TPACK.

4. RESULTS

Table 2. The number of evaluator's agreement related to TPACK

Dimensions	1st evaluation	2nd evaluation
	Agreement	Agreement
CK	4	4
PK	4	4
TK	1	4
PCK	3	4
TPK	1	3
TCK	1	3
TPCK	1	3
C x K	1	4

4.1. The Case Teacher's TPACK Has Improved, Especially in Appropriate Use of Animations and Videos as Part of the Interactive Whiteboard

In terms of the dimensions of TPK, TCK and TPCK, the number of the evaluators' agreement has increased from the first evaluation to the second evaluation (each of the three dimension changed from one agreement to three agreement). The increase showed that the case teacher, Amy, had improved her technology-related knowledge. She further incorporated some dynamic materials, such as animations or videos to express/describe some abstract concepts; also, these dynamic materials could make up for the lack of static instructional materials. Most importantly, she perceived learning IWB was the basic foundation to integrate the IWB technology into pedagogy.

> The first time Amy taught the locations of the digestive tract, she used concrete pictures, but second time she used the IWB to teach the location of the endocrine system. (T2 observation)

> Since the endocrine glands are very abstract, I decided to use animation to explain the function of each part, and the effect was good. (Amy Interview)

> Amy used a video to describe intestinal villi, and used an animation to describe how food enters the throat and esophagus when swallowing; it was very realistic. (R observation)

> I successfully considered how to use the IWB as the base for integrating technology into teaching and learning activities in this lesson. (Amy reflection)

4.2. Observation Stimulates the Science Teacher to Self-Reflect on Their Own Development of TPACK

Both the peer teaching observations provided many comments and suggestions, guiding Amy to reflect upon her own teaching. Amy took those suggestions and spent most of her preparation time to practice how to best integrate IWB technology into her own instructional unit structures and organizations. On the whole, her teaching flow was good; she had further exemplified good-quality teaching knowledge by giving some relevant daily-life examples.

> Amy took the peer advice and had good teaching flow this time, which shows that she prepared well and had good teaching knowledge. (T1 observation)
>
> After the first classroom observation and peer comments, I tried to see how to arrange and organize the instructional materials well, and how to arrange them in the best sequence. (Amy Interview)
>
> In order to have a good teaching observation, I spent most of my preparation time practicing how use IWB technology in conjunction with my teaching. (Amy reflection)
>
> When teaching the digestive system, Amy used the example from daily life that not eating breakfast and stress would cause stomach trouble, but second time she used the IWB to show the picture. (T3 observation)

4.3. The Interactive Whiteboard Can Not Only Enhance Teacher-Student Interaction but Evaluate Student's Knowledge

According to the evaluation result, the four evaluators agree that the case teacher developed both Technology Knowledge and Context Knowledge. This showed that Amy not only used the IWB technology to express the concepts to be taught, but also enhanced the interaction between teachers and students. IWB can further be used to evaluate what students have learned, allowing them to concretely express their knowledge on the whiteboard.

> The first time Amy taught the digestive system, time she used the IWB to explain and teach most of the, but the second time when she taught the endocrine system, the teacher-student interaction had improved. (T2 observation)
>
> First, I conceptually explained the whole endocrine system, and then used the IWB to summarize the overall concept, and finally I used the whiteboard to evaluate what students have learned. (Amy Interview)
>
> Amy would use the IWB to let students answer the locations and functions of endocrine glands in the human body. (T1 observation)

4.4. There Are Still Some Limitations That Need to Be Overcome in Interactive Whiteboard Instruction

On the whole, the results of this instance of classroom observation were good, but there are still some limitations and blind spots in interactive whiteboard instruction, which need to be improved upon in future instruction.

> The projector, as part of the electronic equipment, was not bright enough; the classroom was too dark; and some students wanted to doze off, so this was not beneficial to interactive whiteboard instruction. (T1 observation)
>
> The teacher's presentation words using PowerPoint on the interactive whiteboard were too small. (T2 observation)
>
> When students operated the interactive whiteboard, they used the "drag" function too many times, which shows a problem with the hardware and software. (T3 observation)
>
> The instructional CD-ROM was insufficient in content, and could not conform to the needs of the classroom units since this course requires searching online for necessary items. (Amy Interview)

5. Discussion and Implications

Previous studies on the TPACK scale did not emphasize science and specific units (Graham et al., 2009), and there are few studies emphasizing science teacher TPACK with IWB Technology (Jang, 2010). The current study was based on existing theories, developed an evaluative criteria based on IWB Technology, and redefined the overall TPACK from an integrative view (Angeli & Valanides, 2009). The TPACK theoretical framework established by Mishra and Koehler (2006) includes dimensions such as CK, PK, TK, PCK, TPK, TCK, and TPCK, while the knowledge of TPK, TCK, and TPCK interacting with technology is difficult to clearly define (Angeli & Valanides, 2009). Archambault and Barnett (2010) also reconstructed the theoretical basis, with three categories of PCK, TPCK, and TK, showing that there are clear boundaries only for TK, though the other two are still difficult to clearly delineate. In this study, we find the similar result that all technology-related knowledge (TK TPK, TCK, and TPCK) about the science teacher was improved at the same time. This echoed the findings of technology-related interactions are difficult to separate clearly (Angeli & Valanides, 2009; Archambault & Barnett, 2010). In order to improve the TPACK of the science teacher, technology knowledge was considered as the base for improving the interaction knowledge of TPK, TPK and TPCK through the teaching experience.

In this study, the use of IWBs not only affects innovative teaching (Glover, Miller, Averis & Door, 2007; Higgins, Beauchamp & Miller, 2007; Smith, Higgins, Wall & Miller, 2005), but also develops the TPACK of the science teacher (Holmes, 2009; Jang, 2010). In terms of PCK, the case teacher did now show much difference between the two peer evaluations, but improved greatly in terms of TPCK, explaining the influence of peer instruction observation and evaluation on the case teacher. In peer observation, the observer should act to provide the observee with as an objective view of the teaching session as possible, and review and reflect on that experience with the observee in a way that informs future thinking and practice. (Hammersley-Fletcher & Orsmond, 2004).

Furthermore, the case teacher was able to improve on her teaching through reflecting on the observation, so she incorporated animation and videos in the electronic whiteboard, and improved teacher-student interaction and evaluation of student knowledge. The central goal of reflection is to develop the case teacher's understanding of why she employs certain instructional strategies and notice how they can improve her teaching in order to have a positive effect on students (Lee, 2005). A reflective process can be a developmental activity as well as a means of enhancing the TPACK of the instructor. Furthermore, reflective practice is becoming the favored paradigm for continuing professional development (Clegg, Tan & Saeidi, 2002). On the whole, her TPACK know-how through integration of IWB technology in instruction was obviously improved.

This study attempted to integrate the IWB into the context of science instruction, but some limitations and blind spots of interactive whiteboard instruction were discovered that need to be improved upon in future instruction. In the future, it is possible to use this scale with more science teachers to explore and evaluate the TPACK development. The scale items can also be added to or modified based on different technology types or different subjects.

REFERENCES

Anderson, N. A., Barksdale, M. A. & Hite, C. F. (2005). Preservice teachers' observations of cooperating teachers and peers while participating in an early field experience. *Teacher Education Quarterly*, *32*(4), 97-117.

Angeli, C. & Valanides, N. (2009). Epistemological and methodological issues for the conceptualization, development, and assessment of ICT–TPCK: Advances in technological pedagogical content knowledge (TPCK). *Computers & Education, 52,* 154-168.

Archambault, L. M. & Barnett, J. H. (2010). Revisiting technological pedagogical content knowledge: Exploring the TPACK framework. *Computers & Education*, *55*(4), 1656-1662.

Clegg, S., Tan, J. & Saeidi, S. (2002) Reflecting or acting? Reflective practice and continuing professional development in higher education, *Reflective Practice, 3*(1), 131–146.

Crison, C., Lerman, S. & Winbourne, P. (2007). Mathematics and ICT: A framework for conceptualising secondary school mathematics teachers' classroom practices. *Technology, Pedagogy and Education,* 16(1), 21-39.

Jimoyiannis, A. (2010). Designing and implementing an integrated technological pedagogical science knowledge framework for science teachers professional development. *Computers & Education, 55*, 1259–1269.

Gillen, J., Littleton, K., Twiner, A., Staarman, J. & Mercer, N. (2008). Using the interactive whiteboard to resource continuity and support multimodal teaching in a primary science classroom. *Journal of Computer Assisted Learning, 24*, 348–358.

Glover, D., Miller, D., Averis, D. & Door, V. (2005). The interactive whiteboard: A literature survey. *Technology, Pedagogy and Education,* 14(2), 155-170.

Glover, D., Miller, D., Averis, D. & Door, V. (2007). The evolution of an effective pedagogy for teachers using the interactive whiteboard and modern languages: An empirical analysis from the secondary sectors. *Learning, Media and Technology,* 32(1), 5-20.

Graham, C., Burgoyne, N., Cantrell, P., Smith, L., Clair, L. & Harris, R. (2009). TPACK development in science teaching: Measuring the TPACK confidence of inservice science teachers. *TechTrends, 53*(5), 70-79.

Grossman, P. L. (1990). The making of teacher: Teacher knowledge and teacher education. New York: Teacher College Press.

Hall, I. & Higgins, S. (2005). Primary school students' perceptions of interactive whiteboards. *Journal of Computer Assisted Learning, 21*, 102–117.

Hammersley-Fletcher, L. & Orsmond, P. (2004) Evaluating our peers: Is peer observation a meaningful process? *Studies in Higher Education,* 29(4), 489–503.

Hennessy, S. (2006). Integrating technology into the teaching and learning of school science: A situated perspective on pedagogical issues in research. *Studies in Science Education, 42*(1), 1–48.

Hennessy, S., Deaney, R., Ruthven, K. & Winterbottom, M. (2007). Pedagogical strategies for using the interactive whiteboard to foster learner participation in school science. *Learning, Media and Technology, 32*, 283–301.

Higgins, S., Beauchamp, G. & Miller, D. (2007). Reviewing the literature on interactive whiteboards. *Learning, Media and Technology,* 32(3), 213-225.

Holmes, K. (2009). Planning to teach with digital tools: Introducing the interactive whiteboard to pre-service secondary mathematics teachers. *Australasian Journal of Educational Technology, 25*(3), 351-365.

Jang, S. J. (2008). The effects of integrating technology, observation and writing into a teacher education method course. *Computers & Education, 50*(3), 853-865.

Jang, S. J. (2010). Integrating the interactive whiteboard and peer coaching to develop the TPACK of secondary science teachers. *Computers & Education, 55*(4), 1744-1751.

Jang, S. J. & Chen, K. C. (2010). From PCK to TPACK: Developing a transformative model for pre-service science teachers. *Journal of Science Education and Technology, 19*(6), 553-564.

Koehler, M. J. & Mishra, P. (2005). What happens when teachers design educational technology? The development of Technological Pedagogical Content Knowledge. *Journal of Educational Computing Research, 32*(2), 131-152.

Koehler, M. J., Mishra, P. & Yahya, K. (2007). Tracing the development of teacher knowledge in a design seminar: Integrating content, pedagogy, and technology. *Computers & Education, 49*, 740–762.

Lee, H.-J. (2005). Understanding and assessing preservice teachers' reflective thinking. *Teaching and Teacher Education, 21*, 699–715.

Lopez, O. S. (2010). The digital learning classroom: Improving English language learners' academic success in mathematics and reading using interactive whiteboard technology. *Computers & Education, 54*, 901–915.

Ministry of Education (2008). Whitepaper for educational technology in K-12 education from 2008 to 2011. Taiwan: Author.

Mishra, P. & Koehler, M. J. (2006). Technological pedagogical content knowledge: A new framework for teacher knowledge. *Teachers College Record. 108*(6), 1017-1054.

Niess, M. L. (2005). Preparing teachers to teach science and mathematics with technology: Developing a technology pedagogical content knowledge. *Teaching and Teacher Education, 21*(5), 509-523.

Norris, S. P. (1984). Defining observational competence. *Science Education, 68*(2), 129-142.

Patton, M. Q. (1990). Qualitative evaluation and research methods. Newbury Park, CA. Sage Publication.

Powell, W. & Napoliello, S. (2005). Using observation to improve instruction. *Educational Leadership, 62*(5), 52-55.

Schmid, E. C. (2008). Potential pedagogical benefits and drawbacks of multimedia use in the English language classroom equipped with interactive whiteboard technology. *Computers & Education, 51*, 1553–1568.

Schmidt, D. A., Baran, E., Thompson, A. D., Mishra, P., Koehler, M. J. & Shin, T. S. (2009). Technological pedagogical content knowledge (TPACK): The development and validation of an assessment instrument for preservice teachers. *Journal of Research on Technology in Education, 42*(2), 123-149.

Sedig, K. & Liang, H. (2006). Interactivity of visual mathematical representations: Factors affecting learning and the cognitive process. *Journal of Interactive Learning Research,* 17(2), 179-212.

Shenton, A. & Pagett, L. (2007). From 'bored' to screen: The use of the interactive whiteboard for literacy in six primary classrooms. *Literacy,* 41(3), 129-136.

Shulman, L.S. (1986). Those who understand: Knowledge growth in teaching. *Educational Researcher, 15*, 4–14.

Shulman, L.S. (1987). Knowledge and teaching: Foundations of the new reform. *Harvard Educational Review, 57*, 1–22.

Slay, H., Sieborger, I. & Hodgkinson-Williams, C. (2008). Interactive whiteboards: Real beauty or just lipstick? *Computers & Education, 51*, 1321–1341.

Smith, H. J., Higgins, S., Wall, K. & Miller, J. (2005). Interactive whiteboards: Boon or bandwagon? A critical review of the literature. *Journal of Computer Assisted Learning,* 21, 91-101.

Strauss, A. (1987). *Qualitative analysis for social scientists*. New York: Cambridge University Press.

Tamir. P. (1988). Subject matter and related pedagogical knowledge in teacher education. *Teaching and Teacher Education, 4*, 99-110.

Thompson, A. D. & Mishra, P. (2007). Breaking news: TPCK becomes TPACK! *Journal of Computing in Teacher Education, 24*(2), 38.

Torff, B. & Tirotta, R. (2010). Interactive whiteboards produce small gains in elementary students' self-reported motivation in mathematics. *Computers & Education, 54,* 379-383.

Warwick, P. & Kershner, R. (2008). Primary teachers' understanding of the interactive whiteboard as a tool for children's collaborative learning and knowledge-building. *Learning, Media and Technology, 33*(4), 269–287.

Waxman, H. C. (1995). Classroom observation of effective teaching. In A. C. Ornstein (Ed.), *Teaching: Theory into Practice*. Boston: Allyn and Bacon.

Wellington, J. (2005). Has ICT come of age? Recurring debates on the role of ICT in education, 1982–2004. *Research in Science and Technological Education, 23*, 25–39.

Chapter 7

EXPLORING PEER COACHING ON SCIENCE TEACHERS' TPACK AND STUDENTS' LEARNING OUTCOMES

ABSTRACT

The purpose of this study was to examine how peer coaching affects the Technological Pedagogical and Content Knowledge (TPACK) development of science teachers and student learning outcomes. Peer coaching is socially constructivist in nature because learning depends upon constructing personal knowledge for teaching through social interactions in a community of practice. A peer coaching-based model for TPACK was developed. Three certified science teachers and three classes of eighth graders were assigned to the experimental group, and three science teachers and three similar classes were assigned to the control group. We used a mixed methods design, incorporating both quantitative and qualitative techniques. The main data included students' scores, written assignments, reflective journals and interviews. The results showed that students in the experimental group scored higher than those in the control group, and the two groups had a significant difference in respect to student achievement. The experimental science teachers were clearly aware of students' prior conceptions of the subject matter and their learning difficulties. They also learned to implement technology and multiple teaching methods in order to integrate the subject-matter knowledge. In conclusion, the peer coaching-based model offered science teachers practical opportunities to promote their TPACK, and significantly affected experimental students' learning outcomes. The research implications of this study are provided along with suggestions.

Keywords: Peer coaching; Science teachers; Students' learning outcomes; Technological Pedagogical and Content Knowledge

1. INTRODUCTION

Science teachers in the 21st century need to be equipped with professional teacher knowledge, which has been referred to as Technological Pedagogical and Content Knowledge (TPACK). This implies that teachers have the ability to apply technology into their pedagogical strategies and content representations when teaching specific topics to promote students' learning efficiency (Mishra & Koehler, 2006; Niess, 2005). The TPACK framework builds on Shulman's (1986, 1987) construct of Pedagogical Content Knowledge (PCK) to include technological knowledge as situated within content and pedagogical knowledge. Development and research for science teachers' PCK has received more emphasis (De Jong, Van Driel & Verloop, 2005; Gess-Newsome & Lederman, 1993; Grossman, 1990; Loughran, Mulhall & Berry, 2004; Van Driel, De Jong, & Verloop, 2002). In today's world, it cannot be denied that educational technology is gradually playing a more critical role in our lives, probably because applications of technology to learning environments can be closely linked to students' learning achievement (Kopcha, 2010). Angeli and Valanides (2009) suggested that if teachers can make good and proper use of educational technology, it is possible to create better learning environments for students. Nevertheless, successful incorporations into instruction of educational technology are not that simple. Unlike merely adding a new technology to classroom instructions, it requires teachers to simultaneously have sufficient PCK and technological knowledge to maximize teaching effectiveness and efficiency (Jang, 2010). In line with this, this study focuses on the issues of TPACK to help science teachers achieve more positive and preferable teaching and learning outcomes.

Science teachers should be equipped with the abilities to integrate technology as well as aim for effective teaching. Effective professional growth must be collaborative, involving a sharing of knowledge among teachers' communities of practice (Darling-Hammond & McLaughlin, 1995; Firestone & Rosenblum, 1998). Researchers have reported that regular opportunities for the interaction with colleagues are essential for creating professional school cultures (Lieberman, Saxl, & Miles, 1988; Miller, 1988). A related approach to increase collaboration among teachers is peer coaching. Peer coaching is a collaborative and confidential process through which two or more professional colleagues work together to provide in-class assistance, reflect on current practices, build new skills and knowledge, share ideas, and solve problems (Joyce & Showers, 1995; Slater & Simmons, 2001). Joyce and Showers (1995) suggested that teachers learn from each other in the process of planning instruction, developing the materials to support it, watching each other work with students, and thinking together about the impact of their behavior on the learning process of their students. Previous studies have already indicated that peer coaching can play a critical role in the growth of PCK (Jang, 2009; Lu, 2010), mainly because knowledge acquisition and sharing among peers enables teachers to further reflect on their teaching performance, which in turn helps teachers improve their PCK (Nilsson, 2008). These investigations have examined the practicality of peer coaching for promoting changes in teachers' pedagogical practices and expertise (Jang, 2010; Kohler & Ezell, 1999; Pugach & Johnson, 1995).

Murray, Ma and Mazur (2009) have indicated that although peer coaching seems to be positively related to teachers' collaborative interactions, the effect of peer coaching does not subsequently result in the improvement of student learning. Some studies further indicated that teachers' professional knowledge could be one of the key factors that have a positive effect on student learning (Kanter & Konstantopoulos, 2010; Major & Palmer, 2002; Mansor, Halim & Osman, 2010). For instance, it was found that the growth of teachers' content knowledge and PCK could be positively linked to student learning (Kanter & Konstantopoulos, 2010). Another study demonstrated that a teacher's knowledge could play a positive role in facilitating students' conceptual understandings (Mansor et al., 2010). However, how peer coaching affects the TPACK development of science teachers and students' learning outcomes have not yet been fully examined. Therefore, the purpose of this study was to examine how a developed peer coaching model affects the science teachers' TPACK and the relation between peer coaching and students' learning achievement. In this study, the concepts of "density and buoyancy" were chosen to explore the effects of the intended design since students have particular difficulty understanding abstract, invisible, process and hierarchical level attributes (Murray, Schultz, Brown & Clement, 1990; She, 2005).

2. LITERATURE REVIEW

2.1. TPACK Theoretical Framework

The conceptualization of TPACK is mainly derived from the theoretical framework of PCK. It has been shown that teachers' PCK, containing at least pedagogical knowledge and subject matter knowledge, can be highly associated with teaching effectiveness and efficiency (Shulman, 1986, 1987). Niess (2005) indicated that a teacher's PCK is an integration and interaction of teaching knowledge and subject matter knowledge. Jang and Chen (2010) added

> PCK involves the transformation of other types of knowledge (subject matter knowledge, pedagogical knowledge, and knowledge of context) into viable instruction so that it can be used effectively and flexibly in the communication process between teachers and learners during classroom practice (p. 554).

In other words, PCK not only stresses the important role of a teacher's pedagogical knowledge and subject matter knowledge in instructions, but also emphasizes the teacher's understanding of student learning capabilities and obstacles (Angeli & Valanides, 2009). Although there is no accurate and precise definition of PCK, generally, it is agreed the content of PCK could broadly include "teachers' knowledge of representations of subject matter, and their knowledge of learners' conceptions and content-related difficulties" (Angeli & Valanides, 2009, p.156).

The TPACK proposed by Mishra and Koehler (2006) has been one of the focal points in recent educational technology studies (Archambault & Barnett, 2010; Jang, 2010; Jimoyiann, 2010; Kramarski & Michalsky, 2010). It has been shown that a teacher's technological knowledge is another critical component that could be highly associated with teaching effectiveness and efficiency (Angeli & Valanides, 2009; Chai, Koh & Tsai, 2010). More noteworthy, the initial conceptualization of TPACK has further extrapolated technological content knowledge (TCK), technological pedagogical knowledge (TPK), and technological pedagogical content knowledge (TPCK), which are three potential and additional components that could be closely connected with teaching effectiveness and efficiency. Although the inceptive theoretical framework of TPACK seems promising and attractive, the complex, unclear, and untenable nature of the TPACK framework has remained a tricky problem in recent studies (Angeli & Valanides, 2009; Archambault & Barnett, 2010; Chai et al., 2010).

Koehler, Mishra, and Yahya (2007) used an approach based on discourse analysis to track the development of TPACK. Analyzing the conversations of teachers working in design teams, they tracked the development of each of the seven components of TPACK over the course of a semester. This approach, however, was especially time-consuming and is methodologically specific to the unique context in which it was used (i.e., semester-long design experiences). Angeli and Valanides (2009) explored the use of design-based performance assessments embedded into course sequences. They used self-assessment, peer assessment, and expert assessment of these design-based performances as formative and summative assessments of teachers' understanding. Jang (2010) found that science teachers used IWBs as instructional tools to share their subject matter knowledge and to express students' understanding. In addition, the IWBs helped the science teachers who encountered teaching difficulties in the traditional classroom to better implement their representational repertoires and instructional strategies. Finally, the proposed model of integrating IWBs and peer coaching could develop the TPACK of science teachers.

One recent study by Archambault and Barnett (2010) tested online teachers' perception of TPACK. They revisited and explored the TPACK framework by using factor analysis based on a survey with 24 items designed to measure each of the areas described by the TPACK framework. While the framework is helpful from an organizational standpoint, however, it is difficult to distinguish each of the domains. Chai et al. (2010) addressed 889 pre-service teachers to investigate their perceptions of teachers' TPACK. Although both exploratory factor analysis (EFA) and confirmatory factor analysis (CFA) indicated a robust factor structure with four TPACK components (TK, CK, PK, TPACK), the findings of their study still failed to probe the other potential and underlying components such as TCK and TPK in the initial TPACK framework.

2.2. Peer Coaching

The peer coaching approach is socially constructivist in nature because learning depends upon constructing personal knowledge for teaching through social interactions in a community of practice (Jang, 2007; Vygotsky, 1978). This form of thinking and dialogue among science teachers aligns reflection closely with practice. The role of discourse in shaping what is viewed as legitimate understanding within a scientific community becomes a crucial factor in the construction of knowledge. Thus, to understand how scientific knowledge is constructed, in some part the question becomes one of understanding how knowledge is negotiated or co-constructed in social settings. Co-construction refers to the process of jointly building an understanding, as would be characterized by interactive experiences (Kittleson & Southerland, 2004). In other words, peer science teachers can be encouraged to use discourse to negotiate their way to understanding in their classroom. This notion of inter-subjectivity allows "the meeting of two minds... each operating on the other's ideas, using the back-and-forth discussion to advance his/her own development" (Rogoff, 1990, p. 149). It also allows for joint thinking, problem solving, and the process of decision-making, from which the teachers appropriate new knowledge (Newman, Griffin & Cole, 1989).

Peer coaching provides a community of practice which can be defined as a group of individuals, who share such commonalities as interests, knowledge, resources, experiences, perspectives, behaviors, language, and practices (Barab & Duffy, 2000; Lave & Wenger, 1991). Bowman and McCormick (2000) suggested that through social interaction, active learning evolves and each participant interprets, transforms, and internalizes new knowledge.

Within the framework of peer coaching, such collaborative discussions allow individuals to develop their own perspectives and to model strengths for others. Pierce and Hunsaker (1996) stated that peer coaching not only increases collegiality, but also enhances each teacher's understanding of the concepts and strategies of teaching, and sustains the movement toward restructuring the traditional evaluation efforts by strengthening the ownership of change. Jenkins et al. (2005) suggested peer coaching as a means of developing pedagogical content knowledge because of the real-life context in which teaching and learning occur. Peer coaching can increase reflective practice, aid implementation of teaching models and instructional strategies, and enhance classroom management and development of PCK (Jenkins & Veal, 2002).

Joyce and Showers (1982) introduced peer coaching as a component of in-service teacher training. In their study, the fully elaborated in-service peer coaching model with a planning and implementation focus consisted of four elements: (1) the study of the theoretical basis or rationale of the teaching method; (2) the observation of demonstrations by persons who are experts in the teaching method; (3) practice and feedback in relatively protected conditions; and (4) coaching one another to assist the new method to be incorporated into their day-by-day teaching style. In their more recent work, Joyce and Showers (1995) expanded their view of peer coaching, emphasizing learning through collaborative planning, development and observation of instruction. They stressed the importance of a non-hierarchical relationship between peers working and learning collaboratively to improve their teaching.

Peer coaching, which generally refers to knowledge sharing with peer partners, is often used interchangeably with "peer mentoring, learning centered supervision, peer supervision, and cognitive coaching" in educational studies (Britton & Anderson, 2010, p.306). One of the key advantages of peer coaching in teacher education is that it can provide teachers with a collaborative channel and learning community to share teaching knowledge with each other (Jang, 2010; Britton & Anderson, 2010). For example, Goker (2006) has shown that peer coaching is positively linked to teachers' self-efficacy and growth of instructional skills. Bakkenes, Vermunt, and Wubbels (2010) indicated that peer coaching has a positive influence on the advancement of PCK. Considering the key connection between peer coaching, TPACK and student learning, accordingly, it is critical that peering coaching be incorporated into the theoretical framework of this study.

2.3. Developing a Peer Coaching-Based Model for TPACK

Shulman (1987) proposed that PCK development might pass through the processes of comprehension, transformation, instruction, evaluation, reflection and new understanding. In this study, peer collaboration can be described as a collegial approach to the analysis of teaching aimed at integrating new skills and strategies in classroom practice (Joyce & Showers, 1995). Furthermore, a strong line of recent research outside of traditional science education has begun to use a research approach to professional development (Lumpe, 2007; Marzano, 2003). It is time for science educators to emphasize peer interaction as well as student learning results. Therefore, this study developed a peer coaching-based model by revising Lumpe's (2007) model. The proposed model can be applied to peer collaboration strategies and evaluation of students' results as depicted in Figure 1.

The peer coaching-based model includes two parts: the process of TPACK development and the content of TPACK development. The process of TPACK development involves forming peer-coaching communities and applying the TPACK-RIER (Research, Instruction, Evaluation, and Reflection) through formal training courses/workshops related to four elements. First, this model starts with the study of the theoretical basis or rationale of the specific content teaching method. The research includes an analysis of the topics of textbooks and TPACK articles in teams. The analyses and discussions of these TPACK research articles also contribute to the science teachers' PCK and TPACK of useful instructional strategies, which can help them overcome secondary students' learning difficulties (Jang, 2010; Van Driel, De Jong & Verloop, 2002). Second, because some teachers already had teaching experience, this study applied the instruction implementation in real classrooms and video-recordings, rather than observing expert teachers' teaching observations. Many related studies have indicated that instructional models related to teaching experience are important for PCK and TPACK development (Jang, 2010; Loughran, Mulhall & Berry, 2004; Van Driel, De Jong & Verloop, 2002).

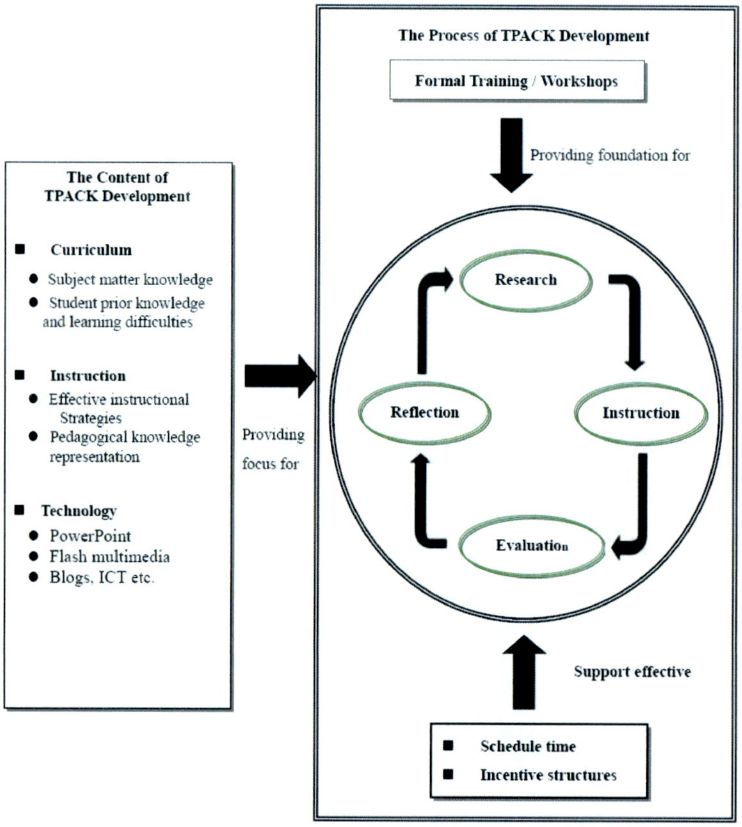

Figure 1. A peer coaching-based model of TPACK development.

Third, peer teachers were evaluated on the basis of students' learning results. Some studies have found that the professional development of teachers might be evaluated by students' learning results or feedbacks (Czerniak, Beltyukova, Struble, Haney & Lumpe, 2006; Murray et al., 2009). DuFour (2005) identified three major ideas that characterize the basis of all professional learning communities, ensuring that students learn, building a culture of collaboration, and focusing on results. Hence, Dufour thinks that not only is peer collaboration conducive to professional learning, but students' learning results may also reflect teachers' professional development. Finally, students' learning results bring out teachers' reflection. Each science teacher was also required to show the videotapes of his/her teaching to share his/her teaching experience with others. This teaching practice can stimulate teachers' self-reflection. To reflect is to think about where one has been and/or what has happened in order to clarify an experience. The act of reflection is primarily concerned with developing insights and discovering solutions to difficulties, or what might be described more correctly as learning opportunities, to revise the instructional strategies of the next topic (Vidmar, 2006). In addition, schedule time and incentive structures must be in place to support the research-based systems.

TPACK highlights the connections and interactions among content, pedagogy, and technology (Mishra & Koehler, 2006). The major content of TPACK development for this model includes three important factors as shown in Figure 1, which are classified into three categories: (1) curriculum: subject-matter knowledge, knowledge of learners including prior knowledge and learning difficulties of learners; (2) instruction: pedagogical knowledge including effective teaching strategies, and knowledge representation; and (3) technology: For instance, PowerPoint is suitable for lecture-type presentation; Flash multimedia is ideal for presenting complex procedures in sequential ways; and Blogs are appropriate for student reflection and social connection.

With knowledge and awareness of technology in mind, teachers are able to rethink course elements that are difficult to teach in traditional ways, and attempt to transform their instruction into better representations using technologies (Angeli & Valanides, 2009; Jang, 2008). On the other hand, for technology to be an integral part of teaching and learning, teachers must also "develop an overarching conception of their subject matter with respect to technology and what it means to teach with technology" (Niess, 2005, p. 510). When content, pedagogy, and technology are well-integrated to facilitate students' knowledge construction, teachers are well-equipped with TPACK, a consolidated knowledge system that promotes students' learning.

3. Methodology

3.1. Participants

In order to avoid any misunderstanding or inefficient communication that might yield bias results, this study found one participant, with whom the researcher could conduct the study and communicate easily. Then, this participant looked for other teachers whom she could trust to become partners for peer coaching. The researcher formed a peer-coaching team with three science teachers who teach eighth-grade students in a secondary school in Taoyuan County, Taiwan. These three science teachers were all enthusiastic about teaching and gaining new knowledge and were willing to make changes. Teacher Mary majored in Biology and has 12 years of teaching experience. She pays a great deal of attention to students' reaction toward her teaching; she, therefore, has the desire to learn new teaching methods and understand students' learning performances. Teacher Amy majored in Physics and has five years of teaching experience. She enjoys improving herself through self-reflection and correcting her own mistakes. She is interested in teaching and creating new teaching methods. She likes discovering and discussing the learning difficulties faced by students. Teacher Dick majored in Earth Science and has taken some courses in Technology. It is his third year of teaching. He enjoys discovering and discussing the feasibility of improving students' learning achievement. One of the classes of each teacher was selected for this study. The numbers of students in each class were 32, 34, and 31. The numbers of students totaled 96. Because the researcher was restricted in that the original class placement of students was done by the school according to a normal S distribution, the subjects were chosen through a non-random sampling. Students at this level were willing to express their own opinions and ask questions concerning the teaching contents. Nevertheless, the learning background of each student had a huge impact on the gap in learning.

The control group in this study included 94 students who were from three different teachers and their eighth-grade classes wherein students had similar academic achievements. Peer-coaching was *not* implemented for the teachers among the three classes, and the control group received a traditional teaching method. The average scores of the two groups(experimental and control) were 70.62 and 70.38, respectively. To ensure that the science grades of the Experimental Group and the Control Group were not statistically different before the experiment, an ANOVA analysis was employed based on the average mean scores of both the control group and the experimental group in the previous semester. The result of the one-way ANOVA ($F=0.472$, $P>.05$) did not show a significant difference. Therefore, we can assume that the experimental and the control groups are equal in terms of science grades.

3.2. Research Design and Implementation

In this study, the Experimental group chose the topic of "density and buoyancy" and used the peer-coaching based TPACK-RIER model (Research, Instruction, Evaluation, and Reflection) to implement the whole process. The research design included one-semester of peer coaching, held once every week. The model's planning and implementation focused on the following four elements.

3.2.1. TPACK research and comprehension

From the beginning of the semester, the researcher made sure team teachers who were going to do the research understood the purpose of this study, the way to implement it, and the resources related to this study. Before teaching, science teachers were required to answer the following questions and write down responses in their reflective journal. The collection of teachers' PCK and TPACK helps teachers better understand their current personal TPACK.

> Assignment 1: From your earlier experiences as a secondary student or from your previous teaching practice, what difficulties have you encountered in learning the concepts of "density and buoyancy"?

The science teachers wrote down their recollections individually, which were then discussed by all. After group discussion, science teachers wrote down students' understandings and preconceptions of these topics in their reflective journals. Then, the researcher assigned a topic, for example, the unit of "density and buoyancy" to let teachers collect and read the relative PCK and TPACK research references and books in a group.

3.2.2. Implementation of Teaching and Video Recording

Every teacher selected appropriate technology, teaching strategy or teaching representation to integrate the instruction design according to the knowledge acquired from the study of relative references or books. Teachers implemented the teaching design in their own classes.

For example, Dick used multimedia VCD to illustrate the different objects of density and their applications in order to stimulate students' motivation. The entire teaching session was video-recorded to serve as a reference for mutual observation, learning and reflection during peer coaching.

> Assignment 2: Give some examples of instructional strategies, representation, and technology that you may use to promote students' understanding of this topic.

Again, all written responses were to be collected, and the subsequent group discussion was to be recorded in their reflective journals.

3.2.3. Evaluating Students' Learning Outcomes

To assess teachers' TPACK status and students' learning outcomes, a test for collecting data concerning students' learning outcome was necessary. The evaluation tool was the first mid-term exam, centering on the topics: "density and buoyancy".

The researcher collected the test scores of the two groups to compare their learning outcomes. This test not only helped teachers understand students' learning achievement and served as a reference for teachers' reflection, but also enabled them to adjust their teaching strategy accordingly.

3.2.4. Group Feedback and Personal Reflection

In this section, the science teachers and the researcher observed each teacher's recorded videotapes and analyzed results of students' tests to facilitate teaching reflection. The researcher and science teachers offered teaching suggestions in order to enhance each teacher's TPACK.

> Assignment 3: "What student difficulties did you identify? Describe your TPACK development after you went through the whole process of the TPACK-RIER module.

All written assignments were collected and, again, the concluding group discussions were recorded in their journals. Furthermore, the reflective stage helped them self-examine their current lesson plan design and provided teaching practice in order to modify future teaching practice. At the same time, the researcher conducted personal interviews with individual teachers to understand each teacher's TPACK.

3.3. Data Collection

To monitor the development of TPACK during this model, data was collected at specific moments that were closely associated with the design of TPACK-RIER. The data collected consisted of: (a) the written assignments of each individual science teacher to the questions and assignments included in the model; (b) the reflective journal written by the science teachers through the overall process of the model; (c) the interviews with the three science teachers, transcribed verbatim; and (d) students' learning scores from the two groups.

The questions of the survey from the interview were designed to gain a deeper understanding of the experimental science teachers' conceptions. The interview was conducted after implementing the model, aimed to reveal the extent of transformation of the science teachers' TPACK. It included the teaching skills or technology, and the understanding of students' prior knowledge. According to the information gathered from the interviews, the researcher intended to: (a) confirm their responses to the written assignments and (b) discern the possible discrepancies of their views written down in reflective journals.

3.4. Data Analysis

This study used both quantitative and qualitative research methods. For the quantitative part, statistical analyses of the learning outcomes of the two groups were carried out. The inductive data analysis employed in this study utilized a qualitative framework that allowed the researcher to build patterns of meaning from the data (McMillan & Schumacher, 2001). Four phases, as described by McMillan and Schumacher, were employed for the analysis of the transcripts: (1) continual discovery throughout the research in order to tentatively identify patterns; (2) categorizing and ordering data; (3) refining patterns through determining the trustworthiness of the data; and (4) synthesizing themes. A constant comparative method was employed to compare the data collected through the written assignments, journals and interviews using the categories generated (Strauss, 1987). The data was first collected, coded, compared and then organized into different categories. Then the data was interpreted according to the categories.

In order to increase the research reliability, the interview data was transcribed based on the interviewers' interpretation. Furthermore, the researcher employed data triangulation to cross-validate the data. During the process of data analysis, the research participants also served as criticizers and suggestion-providers to assist the researcher in revising and analyzing the data.

4. Results and Discussion

4.1. The Science Teachers Were Clearly Aware of Students' Prior Conceptions of the Subject Matter and Their Learning Difficulties

In the first writing assignment, three of the science teachers thought that they might have some misconceptions about the density and buoyancy remaining from their earlier experiences as school-boys or -girls, and as university students, or from their previous teaching practice (Assignment 1). They seemed to have sufficient basic knowledge related to density and buoyancy, however, although they did not necessarily understand the abstract nature of the conceptions and the theoretical formula. The three science teachers stated:

> Hard objects have higher density, and objects with higher density have a larger volume and greater weight. (Mary, written assignment)
>
> Lighter objects will float, and heavier objects will sink. Objects floating on the surface of water have more buoyancy than those sinking in the water. (Amy, written assignment)
>
> Since density, mass and volume are abstract conceptions which are complicated to comprehend, students might confuse the mass and the weight in the formulation of density conceptions. (Dick, written assignment)

However, in the third written assignment, "what student difficulties did you identify?", Science teachers considered themselves to be clearly aware of students' prior conceptions of the subject matter and their learning difficulties. Mary understood students' levels of comprehension for the concept of density after teaching practice. Dick used some examples adopted from real-life experience to motivate students to learn. In addition, Amy acquired the ability to integrate theoretical concepts and knowledge into her teaching practices.

> After the teaching practice, it was easier for me to understand students' levels of comprehension for the concept of density. (Mary, reflective journal)
>
> I used some concrete objects as the material for theoretical implications as part of my teaching content. This was intended to reduce confusion due to abstract conceptions, and also to stimulate students' interest in learning. (Dick, written assignment)
>
> I found that some students probably understand the concept of buoyancy, but it was still difficult for them to apply it correctly when solving problems. They did not know how to use a formula to express the physical significance to solve problems. In my opinion, there is a significant difference between students' acquired knowledge in the science course and their problem-solving capability. (Amy, interview)

4.2. The Science Teachers Learned to Implement Technology and Multiple Teaching Methods in Order to Integrate the Subject-Matter Knowledge

After focusing on their expected learning difficulties, science teachers wrote down and discussed instructional strategies that they considered potentially useful to enhance students' understanding (Assignment 2). The main difficulty they encountered was how to present the science formula and content knowledge to students in an efficient way. However, from the TPACK training, the science teachers learned how to implement technology, and multiple teaching methods and strategies of integrating subject-matter knowledge into real-life examples, helping them better explain the content knowledge. The science teachers thought that they might explain important concepts by designing some activities to give students the chance to practice and enrich their empirical experience. Peer coaching provided the opportunities for collaborative teachers to develop their skills in applying technology to design lessons as well as to improve their teaching skills. The three science teachers stated:

> To design a teaching activity for explaining the abstract buoyancy concept formula to students was not easy. Thus, I used materials from everyday life to guide the students. For example, I used clay as a demonstration material, broke it into lumps of different sizes, and put them into water so students could observe whether they floated or sank. (Amy, written assignment)
>
> I think that using the computer when teaching not only helps explore scientific concepts, but also has an entertaining effect, thus, enhancing the students' motivation to learn. I learned many computer skills from the peer coaching in this study. (Dick, reflective journal)
>
> Through peer coaching, we could work collaboratively to elaborate each other's specialties and develop our skills in employing technology to design more interesting and lively teaching contents. I think it will help my teaching. (Mary, interview)

4.3. The Teaching Model Offered Science Teachers' Practical Opportunities to Promote Their TPACK

After focusing on instructional strategies, science teachers wrote down the effect of the TPACK-RIER model that they considered potentially useful to enhance their TPACK (Assignment 3). The teaching model offered science teachers a teaching practice opportunity to promote their TPACK, and so it was possible for the science teachers to connect their professional knowledge of the subject matter with their teaching methods. In addition, the science teachers learned how to interpret the teaching units in a way that was more comprehensive than the traditional teaching method. The science teacher also learned some multimedia technology to illustrate the different objects of density and their applications in class. The three science teachers stated:

> The model gave me lots of practical experience and helped me integrate content knowledge and teaching skills in actual classroom practice. It helped me clarify the areas of strength and weakness in my teaching. (Amy, written assignment)
>
> The model helped me select a proper teaching method to interpret the subject matter for the unit of density in a more comprehensible way, and thereby revised my previous teaching method. (Mary, reflective journal)
>
> I created a multimedia VCD to illustrate the different objects of density and their applications. This idea was inspired through the peer coaching process. (Dick, Interview)

4.4. Effects of Peer Coaching on Students' Learning Outcomes

In order to check if peer coaching really affects the students' learning outcomes, the researcher first adopted an ANOVA analysis to examine the original state of the total mean scores of the two groups. It revealed that students' science achievement was not different between the two groups. To further reduce external confounding that may threaten the validity of the experiment, the researcher adopted the eighth-graders' total mean grades from the previous semester as the covariate and the first mid-term score of the second semester as the dependent variable. We then used ANCOVA to investigate the effect of peer coaching on eighth graders' academic achievement in science. Before the analysis of ANCOVA, the test for homogeneity was evaluated and no significant result was found.

The result of the ANCOVA analysis is shown in table 1. It shows that students in the experimental group scored higher than those in the control group, and there was a significant difference between the two groups ($F= 8.231$, $P< .05$). This indicated that peer coaching could have a positive effect on experimental students' science academic achievement. In other words, the TPACK-RIER model did enhance science achievements of students whose teachers took part in peer coaching.

Table 1. ANCOVA for Mean Scores of Students in the Experimental and Control Groups

Group	N	M	SD	F
E	97	75.85	10.50	8.231*
C	94	71.24	11.10	

Note. Covariate is the mean score (70.50) of students' science grades from the previous semester. *$p < .05$.

5. Discussion and Implications

This study first examines the content of science teachers' initial TPACK of learning difficulties. Initially, the science teachers thought that they might have some misconceptions about density and buoyancy from their earlier experiences, however, the science teachers considered themselves clearly aware of students' prior conceptions of "density and buoyancy" and their learning difficulties after peer coaching practice. By getting acquainted with students' specific conceptions and ways of learning, science teachers might use some examples adopted from real life and practical experience to motivate students to learn. This would restructure their subject-matter knowledge into a more easily understood form that enables productive communication with their students (Lederman, Gess-Newsome & Latz, 1994). This study also provides empirical evidence showing that the teaching model did have some impact on science teachers' TPACK. Since the science concepts and theories learned by the traditional teaching method were usually considered objective and abstract, the science teachers found it easier for them to combine the theory with practice and further organize their subject-matter knowledge through the teaching model. The science teacher also learned some multimedia technology to illustrate the different objects of density and their applications. They developed their TPACK, which has been described as the transformation of several types of knowledge for teaching (Magnusson, Krajcik & Borko, 1999). Following the findings of Geddis (1993), the transformation turned the science teachers' subject-matter knowledge into teachable content knowledge.

Moreover, the science teachers' knowledge of representations and teaching strategies benefited from the actual teaching experience. This strong impact of teaching experiences is consistent with the findings of other scholars (De Jong, Van Driel & Verloop, 2005; Grossman, 1990; Jang, 2010). It was found that the proposed TPACK-RIER model can help science teachers develop technology, multiple teaching methods and strategies for integrating the subject-matter knowledge into real-life examples in the science lessons. When science instructors are well-informed of students' prior knowledge and existing conceptions about the topics to be taught, they are able to determine students' readiness to learn, and set appropriate goals and materials for their students. Then, science instructors must be knowledgeable about the topics students find most inaccessible, the kinds of errors that students commonly make, and then be able to modify instructional representations and strategies for their students (Magnusson et al., 1999; Park & Oliver, 2008).

The science teachers in this study were trained to play two different kinds of roles, collaborative coaching and reflective coaching, in the activities. As collaborative coaches, science teachers were asked to observe and analyze the teaching practice. They learned from each other about how to design more interesting and lively teaching contents, and refine teaching skills through immediate feedback (Bowman & McCormick, 2000; Murray, Ma & Mazur, 2009). Science teachers are also the most appropriate reflective coaches for each other (Vidmar, 2006). In this regard, when the science teachers provided feedbacks to each other, they not only gave suggestive but also positive feedbacks in order to facilitate the teaching skills and strategies of their peers. They should be more open to critical suggestions for TPACK improvement and changes offered by the supportive peers. Therefore, the experimental TPACK-RIER model that includes peer coaching has significant advantages in these respects.

Finally, this study found that peer coaching affects students' learning outcomes. Based on the mid-term scores of the two groups,, there was a significant difference between the peer coaching group and the non-peer coaching group. Scholars agree that successfully applying technology to instruction is closely linked to teaching effectiveness and efficiency (Jang, 2010). However, if science teachers would like to maximize learning their outcome, it is critical that instructional strategies and technology should be centralized on understandings of students' learning capabilities and difficulties. Indeed, it is likely that peer coaching helps science teachers elaborate technology instruction to draw students' attention to the learning content in the beginning. Nevertheless, without sufficient understanding of students' learning capabilities and obstacles, as well as supports for the subject matter knowledge, it will be difficult to help students really acquire new knowledge, skills and the ability to further improve their learning outcomes.

The results of this study show that peer coaching could develop science teachers' TPACK and directly affect students' learning outcomes. This is somewhat congruent with the findings of the previous study. Students' learning outcomes could still rely more on a teacher's PCK (Kanter & Konstantopoulos, 2010). In other words, PCK or TPACK should always be focal points in future educational studies with the primary goal to help students acquire knowledge and improve their learning outcomes. This implies that knowledge of assessment plays a key role in assessing students' understanding, for it provides information about students' prior knowledge, misconceptions, level of mastery, and learning difficulties. Henze, van Driel and Verloop (2008) found science teachers used various ways to assess students' understanding, including written tests, oral and poster presentations, student reflections, classroom debates, and portfolios.

This TPACK study has already made a good attempt to draw people's attention to the TPACK framework through peer coaching. Overall, the study findings add to the body of knowledge in the educational field. This study has not only probed into the relationship between peer coaching and science teachers' TPACK development, but also revealed the influence of peer coaching on students' learning outcomes. This study was conducted to examine science teaching in six eighth-grade classes. Furthermore, the six classes in this study were chosen based which classes were better (experimental group and control group). Therefore, the notion of generalization was not taken into considered in this study. The data also shows that the learning outcomes of this model, in terms of TPACK development, were different for different science teachers, and mostly limited to the specific topics (density and buoyancy) that were focused on. A follow-up study is required to determine whether the design of this model is suitable for other science topics. Again, the innovative TPACK-RIER model provides a way to develop science teachers' TPACK and also serves as a useful reference for other pedagogical goals.

References

Angeli, C. & Valanides, N. (2009). Epistemological and methodological issues for the conceptualization, development, and assessment of ICT–TPCK: Advances in technological pedagogical content knowledge (TPCK). *Computers & Education, 52*(1), 154-168.

Archambault, L. M. & Barnett, J. H. (2010). Revisiting technological pedagogical content knowledge: Exploring the TPACK framework. *Computers & Education, 55*(4), 1656-1662.

Bakkenes, I., Vermunt, J. D. & Wubbels, T. (2010). Teacher learning in the context of educational innovation: Learning activities and learning outcomes of experienced teachers. *Learning and Instruction, 20*(6), 533-548.

Barab, S.A. & Duffy, T.M. (2000). From practice fields to communities of practice. In D.H. Jonassen & S.M. Land (Eds.), *Theoretical foundations of learning environments* (pp. 25–55). Mahwah, NJ: Lawrence Erlbaum Associates.

Bowman, C. & McCormick, S. (2000) Comparison of peer coaching versus traditional supervision effects, *The Journal of Educational Research, 93*(4), 256-264.

Britton, L. R. & Anderson, K. A. (2010). Peer coaching and pre-service teachers: Examining an underutilised concept. *Teaching and Teacher Education, 26*(2), 306-314.

Chai, C. S., Koh, J. H. L. & Tsai, C.-C. (2010). Facilitating pre-service teachers' development of Technological, Pedagogical, and Content Knowledge (TPACK). Educational Technology & Society, 13 (4), 63–73.

Czerniak, C. M., Beltyukova, S, Struble, J., Haney, J. J. & Lumpe, A. T. (2006). Do you see what I see? The relationship between a professional development model and student achievement. In R. E. Yager (Ed.), *Exemplary science in grades 5–8: Standards-based success stories* (pp. 13–43). Arlington, VA: NSTA Press.

Darling-Hammond, L. & McLaughlin, M.W. (1995). Policies that support professional development in an era of reform. *Phi Delta Kappan, 76*, 597-604.

De Jong, O., Van Driel, J.H. & Verloop, N. (2005). Preservice teachers' pedagogical content knowledge of using particle models in teaching chemistry. *Journal of Research in Science Teaching, 42*(8), 947–964.

DuFour, R. (2005). What is a professional learning community? In R. DuFour, R. Eaker, & R. DuFour (Eds.). *On common ground* (pp. 31–43). Bloomington, IN: Solution Tree.

Firestone, W.A. & Rosenblum, S.(1988). Building commitment in urban high schools. *Educational Evaluation and Policy Analysis, 10*(4), 285-299.

Geddis, A. N. (1993). Transforming subject-matter knowledge: The role of pedagogical content knowledge in learning to reflect on teaching. *International Journal of Science Education, 15*, 673–683.

Gess-Newsome, J. & Lederman, N.G. (1993). Preservice biological teachers' knowledge structures as a function of professional teacher education: A year-long assessment. *Science Teacher Education, 77*(1), 25-43.

Goker, S. D. (2006).Impact of peer coaching on self-efficacy and instructional skills in TEFL teacher education. *System, 34*(2), 239-254.

Grossman, P. (1990). The making of a teacher: Teacher knowledge and teacher education. New York: Teachers College Press.

Henze, I., van Driel, J. & Verloop, N. (2008). Development of experienced science teachers' pedagogical content knowledge of models of the solar system and the universe. *International Journal of Science Education, 30*(10), 1321-1342.

Jang, S. J. & Chen, K. C. (2010). From PCK to TPACK: Developing a transformative model for pre-service science teachers. *Journal of Science Education and Technology, 19*(6), 553-564.

Jang, S. J. (2007). A study of the students' construction of science knowledge: Talk and writing in a collaborative group. *Educational Research, 49*(1), 65-81.

Jang, S. J. (2008). Innovations in science teacher education: Effects of integrating technology and team-teaching strategies. *Computers & Education*, 51(2), 646-659.

Jang, S. J. (2009). *PCK and teaching innovations.* NY: Nova Science Publishers, Inc.

Jang, S. J. (2010). Integrating the interactive whiteboard and peer coaching to develop the TPACK of secondary science teachers. *Computers & Education, 55*(4), 1744-1751.

Jenkins, J. & Veal, M.L. (2002). Preservice teachers' PCK development during peer coaching. *Journal of Teaching in Physical Education, 22*(1), 49-68.

Jenkins, J., Garn, A. & Jenkins, P. (2005). Preservice teacher observations in peer coaching. *Journal of Teaching in Physical Education, 24*, 2-23.

Jimoyiann, A. (2010). Designing and implementing an integrated technological pedagogical science knowledge framework for science teachers' professional development. *Computers & Education, 55*(3), 1259-1269.

Joyce, B. R. & Showers, B. (1982). The coaching of teaching. *Educational Leadership, 40*, 4-10.

Joyce, B. R. & Showers, B. (1995). Student achievement through staff development: Fundamentals of school renewal (2nd edition.). White Plains, NY: Longman.

Kanter. D. E. & Konstantopoulos, S. (2010). The impact of a project-based science curriculum on minority student achievement, attitudes, and careers: The effects of teacher content and pedagogical content knowledge and inquiry-based practices. *Science Education, 94*(5), 855-887.

Kittleson, J. M. & Southerland, S. A. (2004). The role of discourse in group knowledge construction: A case study of engineering students. *Journal of Research in Science Teaching, 41*(3), 267–293.

Koehler, M. J., Mishra, P. & Yahya, K. (2007). Tracing the development of teacher knowledge in a design seminar: Integrating content, pedagogy, and technology. *Computers & Education, 49*, 740–762.

Kohler, F.W. & Ezell, H.K. (1999). Promoting changes in teachers' conduct of student pair activities: An examination of reciprocal peer coaching. *Journal of Special Education, 33*(3), 154-165.

Kopcha, T. J. (2010). A systems-based approach to technology integration using mentoring and communities of practice. *Educational Technology Research & Development, 58* (2), 175-190.

Kramarski, B. & Michalsky, T. (2010). Preparing preservice teachers for self-regulated learning in the context of technological pedagogical content knowledge. *Learning and Instruction, 20*(5), 434-447.

Lave, J. & Wenger, E. (1991). Situated learning: Legitimate peripheral participation. Cambridge, UK: Cambridge University Publishers.

Lederman, N.G., Gess-Newsome, J. & Latz, M.S. (1994). The nature and development of preservice science teachers' conceptions of subject matter and pedagogy. *Journal of Research in Science Teaching, 31*, 129–146.

Lieberman, A., Saxl, E.R. & Miles, M.B.(1988). Teaching leadership: Ideology and practice. In A. Lieberman (Ed.) , *Building a professional culture in schools*. New York: Teachers College Press.

Loughran, J.J., Mulhall, P. & Berry, A. (2004). In search of pedagogical content knowledge in science: Developing ways of articulating and documenting professional practice. *Journal of Research in Science Teaching, 41*, 370–391.

Lu, H-L. (2010). Research on peer coaching in preservice teacher education – A review of literature. *Teaching and Teacher Education, 26*(4), 748-753.

Lumpe, A. T. (2007). Research-based professional development: Teachers engaged in professional learning communities. *Journal of Science Teacher Education, 18*, 125–128.

Magnusson, S., Krajcik, J. & Borko, H. (1999). Nature, sources, and development of pedagogical content knowledge for science teaching. In J. Gess-Newsome & N. Lederman (Eds.), *Examining pedagogical content knowledge* (pp. 95-132). Netherlands: Kluwer.

Major, C. & Palmer, B. (2002). Faculty knowledge of influences on student learning. *Peabody Journal of Education, 77*(3), 138-162.

Mansor, R., Halim, L. & Osman, K. (2010). Teachers' knowledge that promote students' conceptual understanding. *Procedia Social and Behavioral Sciences, 9*, 1835-1839.

Marzano, R. J. (2003). *What works in schools: Translating research into action*. Alexandria, VA: Association for Supervision and Curriculum Development.

McMillan, J. & Schumacher, S. (2001). *Research in education: A conceptual introduction (5th ed.)*. New York: Addison Wesley Longman.

Miller, L. (1988) . Unlikely beginnings: The district office as a starting point for developing a professional culture for teaching. In A. Lieberman (Ed.), *Building a professional culture in schools*. New York : Longman.

Mishra, P. & Koehler, M. J. (2006). Technological pedagogical content knowledge: A framework for teacher knowledge. *Teachers College Record, 108*(6), 1017-1054.

Murray, T., Schultz, K., Brown, D. & Clement, J. (1990). An analogy-based computer tutor for remediating physics misconceptions. *Interactive Learning Environment, 1*, 79-101.

Murry, S., Ma, X. & Mazur, J. (2009). Effects of peer coaching on teachers' collaborative interactions and students' mathematics achievement. *Journal of Educational Research, 102*(3), 203-212.

Newman, D., Griffin, P. & Cole, M. (1989). The construction zone: Working for cognitive change in school. Cambridge: Cambridge University Press.

Niess, M. L. (2005). Preparing teachers to teach science and mathematics with technology: Developing a technology pedagogical content knowledge. *Teaching and Teacher Education, 21*(5), 509-523.

Nilsson, P. (2008). Teaching for understanding: The complex nature of pedagogical content knowledge in pre-service education. *International Journal of Science Education, 30*(10), 1281-1299.

Park, S. & Oliver, J. S. (2008). Revisiting the conceptualization of pedagogical content knowledge (PCK): PCK as a conceptual tool to understand teachers as professionals. *Research in Science Education, 38*, 261–284.

Pierce, D. & Hunsaker, T. W. (1996) Professional development for the teacher, of the teacher, and by the teacher, *Education, 117*(1), 101-105.

Pugach, M. & Johnson, L.J.(1995). Unlocking expertise among classroom teachers through structures dialogue: Extending research on peer collaboration. *Exceptional Children, 62*, 101-110.

Rogoff, B. (1990). Apprenticeship in thinking: Cognitive development in social context. New York: Oxford University Press.

She, H.C. (2005). Enhancing eight grade students' learning of buoyancy: The interaction of teachers' instructional approach and students' learning preference styles. *International Journal of Science and Mathematics Education, 3*(4), 609-624.

Shulman, L. S. (1986). Those who understand: knowledge growth in teaching. *Educational Researcher, 15*, 4–14.

Shulman, L. S. (1987). Knowledge and teaching: foundations of the new reform. *Harvard Educational Review, 57*, 1–22.

Slater, C.L. & Simmons, D.L. (2001). The design and implementation of a peer coaching program. *American Secondary Education, 29*(3), 67-76.

Strauss, A. (1987). *Qualitative analysis for social scientists*. New York: Cambridge University Press.

Van Driel, J.H., De Jong, O. &Verloop, N. (2002). The development of preservice chemistry teachers' PCK. *Science Education, 86*, 572–590.

Vidmar, D. J. (2006). Reflective peer coaching: Crafting collaborative self-assessment in teaching. *Research Strategies, 20*, 135-148.

Vygotsky, L.S. (1978). Mind in society. Cambridge, MA: Harvard University Press.

Chapter 8

DEVELOPING THE TPACK OF ELEMENTARY MATHEMATICS TEACHERS IN A PEER COMMUNITY USING INTERACTIVE WHITEBOARD TECHNOLOGY

ABSTRACT

The notion of teachers' pedagogical professional knowledge has increasingly been emphasized in recent years, especially with Technological Pedagogical and Content Knowledge (TPACK). A collaborative action research was conducted by employing a spiral process including planning, implementing, observing, reflecting, and evaluating. The full implementation of IWB took approximately one semester consisting of two action circles. The participants formed a peer community group that contained five elementary teachers from the same school and met once every two weeks. Mathematics topics included counting, position, sequence, time and so on. Data sources of the study contained teaching reflection notes, in-class videotapes, peer community meetings and the TPCK evaluation questionnaire.

The results showed that the characteristics of IWB and the role/position of IWB in math class were highlighted through "learning by doing" in the first action circle. When the teaching content and representation was centered with the using the IWB, the IWB limited the mathematics teaching. The results of the TPACK evaluation questionnaire revealed that there was a lack of student-centered teaching strategies. In this case study, the teacher tended to lectures instead of having more interactions. In accordance with the feedback from circle one, the teacher changed the focus from the center of "Technology" to the center of "PCK" in circle two. Meanwhile, IWB had turned into a teaching tool, when the application of technology was evaluated based on the mathematics teaching content and strategies. The teacher helped students understand the time by authentically recording a student's daily schedule. To enhance students' interactions, the class was divided into small groups with face-to-face seating arrangements for efficient discussions. The teacher served as a facilitator, and her teaching strategies in circle two became more diverse and vivid compared with the ones in circle one (i.e., a teacher-centered lectures).

Although Koehler and Mishra (2005) proposed the TPCK framework with its three interrelated knowledge, Teaching (T), Pedagogy (P) and Content (C), empirically documented studies on mathematics teachers' TPACK with specific content knowledge are scant. This study examined a mathematics teacher's process of developing TPACK through aactive research with two circles. Along with the circular feedbacks, the results of the study also uncovered the limitations of technology. More specifically, when the

IWB was overly emphasized, the transformation of teaching content was restricted. Thus, it is suggested for future work to integrate other tools to fully develop mathematics teachers' TPACK.

Keywords: TPACK; Elementary mathematics teachers; Peer community; Interactive whiteboard technology

INTRODUCTION

The Primary schools have progressively installed the interactive whiteboard (IWB) and other related equipment since the Taiwan Ministry of Education had been promoted the country's e-learning environment in 2006. This new pedagogical tool in educational settings impacts on stimulating students' active participation, promoting interactions between teachers and students and creating positive learning environments in class. When teachers use new technology in teaching, they encounter some difficulties. Thus, teachers can form in groups (Community of teachers) rather than teaching or designing curriculum individually. Teachers in a peer community group can regularly observe, discuss and share ideas together. Bowman and McCormick (2000) suggested that through social interactions, active learning occurs, and each participant interprets, transforms, and internalizes new knowledge.

Shulman (1987) indicated that teachers' professional development is dependent on their understanding of the pedagogical content knowledge (PCK) of a certain subject. PCK addresses how teachers teach through pedagogical strategies such as using analogy and authentic activities to help students conceptualize the related subject contents for improving students' meaningful understandings and helping their learning difficulties (Grossman, 1990). Niess (2005) pointed out that technology is an important element in a student's learning. Teachers integrating appropriate technology in teaching help improve students' comprehension and abilities in learning. TPACK (Technology Pedagogical and Content Knowledge) refers to teachers integrating technology and information into teaching the subject content and promoting student's learning efficiency. It is proposed as the knowledge that includes technology, pedagogy, and content, as well as the knowledge for teaching. Teachers' TPACK progressively develops through the interactions between teachers and students in class and through their teaching practices by using technologies (Angeli & Valandies, 2009). The characteristics of IWBs are in accord with students' learning styles that IWBs can present visual aids such as figures, pictures and animations to help students learn and allow teachers to use various forms of representations to explain mathematical concepts and to propose more creative and effective ways in teaching.

Several studies have further suggested the benefits of using the Interactive Whiteboard (IWB) in classroom settings. They have considered the role of IWBs to increase students' motivation (Glover, Miller, Averis & Door, 2007; Hall & Higgins, 2005; Hennessy, Deaney, Ruthven, & Winterbottom, 2007; Schmid, 2008; Torff & Tirotta, 2010) and student performance (Lopez, 2010; Sedig & Liang, 2006). However, it would appear the means for training teachers to use the IWB and developing the TPACK of mathematic teachers have yet to be fully investigated. Therefore, this study was an action research that intended to investigate an elementary mathematics teacher's TPACK development using IWB technology in a peer community group through a spiral research process of planning, implementating, observing, reflecting, and evaluating.

LITERATURE REVIEW

Interactive Whiteboard Use

Blackboards are an essential tool in the classroom and have been used to display teaching contents and organize teaching activities. However, the drawbacks of blackboards are apparent in that teachers waste time in repeatedly writing and erasing teaching contents on the blackboards, hindering teaching effectiveness. Additionally, blackboards produce chalk ashes that are harmful for people's health. It has been years that multimedia systems have been identified to supplement the drawbacks of blackboard use, and IWBs are becoming one of the technical tools to create a teaching environment where teachers can easily display digital information and integrate this technological device with various functions into their teaching.

Researchers interested in IWB use have conducted studies on examining the integrated use of IWBs and its impacts on teachers' pedagogical strategies (Glover, Miller, Averis & Door, 2005; Higgins, Beauchamp & Miller, 2007; Holmes, 2009; Jang, 2010; Shenton & Pagett, 2007; Smith, Higgins, Wall & Miller, 2005). Smith, Hardman and Higgins (2006), in their study, concluded that classroom discourse produced in the IWB-use classrooms showed more open-ended questions and received more answers from students than the discourse produced in non-IWB use classrooms. In this IWB-use learning environment, the paces of both teaching and learning were enhanced. Other researchers found various advantages associated with IWB use (Gillen, Littleton, Twiner, Staarman & Mercer, 2008; Hall & Higgins, 2005; Kennewell, 2001; Miller & Glover, 2002; Smith et al., 2005; Walker, 2003; Wood, 2001):

1) Promotes teaching efficiency
   The use of IWBs can extend teachers' pedagogical strategies of lecturing and writing on the board simultaneously through visually representing digitalized contents. Some of the benefits of integrating such technology into classroom instruction include: impacts on teachers' classroom management skills; an increase of using multimedia resources such as words, pictures, illustrations, sound data and websites (Smith et al., 2005); and an increase of teaching pace and time (Gillen et al., 2008; Shenton & Pagett, 2007; Walker, 2003). For example, IWBs can help math teachers to display diagrams in a quick way to increase teaching effectiveness. In addition, information shown on the board can be stored and retrieved for latter reviews (Gillen et al., 2008).
2) Reduces teachers' teaching load
   IWBs not only present various forms of visual materials including functions such as moving, making footnotes, enlarging and reducing files and playing animated videos, but also provide functions such as auto-recording and editing their own classes to reduce teachers' load in preparing teaching materials by re-using the teaching materials (Kennewell, 2001).

3) Makes teaching become flexible, accommodative and diverse
   IWBs bring positive impacts in teaching including flexibility and interchangeability such as taking notes, flipping pages, saving files, moving and presenting key points on the board (Hall & Higgins, 2005; Miller & Glover, 2002; Wood, 2001). IWBs serve as a new interface to integrate teaching sources by combing a variety of technologies or teaching software suitable to teaching that can provide students with rich information to get students' attention, help them comprehend abstract concepts, and reinforce reflective memory.
4) Increases classroom interactions
   Smith et al. (2005) indicate that the primary advantage of the IWB is to increase interactions between teachers and students. Teachers should pay attention to the functions of interactivity so they can make IWBs become an effective technology supplemental tool in promoting students' learning (Gillen et al., 2008). Teachers need to integrate various types of interactivities to achieve both teaching effectiveness and learning goals as IWBs can help students internally process the gained information and receive immediate feedback by interacting with multimedia, simulations, and modeling provided by teachers through this particular technology (Lopez, 2010).

The use of IWBs has been studied with pre-service teachers (Holmes, 2009) and in-service teachers in mathematics (Miller et al., 2005) as well as students' understanding of mathematical concepts (Taylor, Harlow, & Forret, 2010). Holmes (2009) examined 13 pre-service mathematics teachers' development of integrating IWBs features in mathematical lesson activities. The participants answered an open-ended question to indicate their opinions for the use of technology in mathematics. They were required to select a mathematical concept to design a mathematical lesson activity by applying the pedagogical theories they learned in the course and integrating IWB features. During the process of designing a mathematical concept with IWB use, pre-service teachers recognized the benefits that IWBs could increase the visualization of presentations, interactivity, and various ways of representing mathematical ideas in order to enhance conceptual understanding.

Moreover, teachers can use IWBs to enhance the relations between cognitive and metacognitive activities and students' reactions to classroom tasks as well as helping students stay engaged (Warwick & Kershner, 2008). IWBs also help capture and hold learners' attention and lead to increased students' motivation (Shenton & Pagett, 2007). Miller, Glover and Averis (2005) studied secondary mathematics teachers' pedagogical development with the use of IWBs. The primary data sources were video-taped lessons, observations and interviews. The results indicated that mathematics teachers gradually and progressively developed the integrative skills of IWBs into their teaching. At the early stage of developing integrative skills of IWBs, they basically used IWBs as only a visual support to display PowerPoint, Excel, or computer programs and lacked practices in using significant features of IWBs into teaching. At the latter stage, teachers indicated advanced understanding of the use of IWBs. The interchangeable process occurred through the interactivity between teachers and students in which students gained mathematical concept knowledge and teachers developed pedagogical skills of integrating IWBs to teaching mathematical lessons. The use of IWBs helped teachers to enhance classroom discussions, explain mathematical processes, explore various representations of mathematics concepts, and develop mathematical hypotheses. In this learning environment, students' cognition was developed as one of the aims designed for the mathematical lessons by teachers. The use of the IWB allowed teachers to be aware of students' preferred learning styles, represent various visual concepts, plan lessons for students' cognitive development and active thinking, provide immediate feedback, and improve recall skills. Therefore, students tend to attain positive outcomes when teachers incorporate the utilization of the IWB features into their current pedagogical strategies and the degree of students' positive learning outcomes is dependent on different teachers (Glover et al., 2007).

Glover et al. (2005) pointed out that one of the advantages of interactive electronic whiteboards is not only to increase interactions between teachers and students, but also interactions among students. When teachers use functions of the IWB such as drag and draw in teaching, students can observe how teachers use these functions. Students can also observe how their peers interact with the whiteboard and develop their skills in using this technological device. Students can interact with each other to activate their conceptual understanding by solving a problem together on the whiteboard.

Peer Coaching

Peer coaching provides a community of practice defined as a group of individuals, who share commonalities such as interests, knowledge, resources, experiences, perspectives, behaviors, language, and practices (Lave & Wenger, 1991). Within the framework of peer coaching, such collaborative discussions allow individuals to develop their own perspectives and to model strengths from other peers. Pierce and Hunsaker (1996) stated that peer coaching not only increases collegiality, but also enhances each other's teacher's understanding of the concepts and strategies related to teaching, and sustains the movement toward restructuring the traditional evaluation efforts by strengthening the ownership of change. Jenkins, Garn and Jenkins (2005) suggested peer coaching as a means of developing PCK because the approach is situated in the context wherein teaching and learning occur. Peer coaching can increase reflective practice, aid implementation of teaching models and instructional strategies, and enhance the classroom management and development of PCK (Jenkins & Veal, 2002; McAllister & Neubert, 1995).

Technological Pedagogical and Content Knowledge

Niess et al. (2009) proposed the process of TPACK development for mathematical teachers based on a TPACK model including five stages: recognizing, accepting, adapting, exploring and advancing.

1) Recognizing: teachers start to use the technological tool in teaching mathematics, but they are unable to integrate technologies in teaching and learning mathematics in classrooms.
2) Accepting: teachers hold positive or negative attitudes in integrating technology with mathematics classes.

3) Adapting: teachers either accept or reject using appropriate technologies in teaching and learning mathematics in classrooms.
4) Exploring: teachers actively integrate technologies into teaching and learning mathematics in classrooms.
5) Advancing: teachers are able to evaluate and reflect on the results of teaching through integrating technologies in teaching and learning mathematics in classrooms.

Koehler, Mishra and Yahya (2007) used an approach based on discourse analysis to track the development of TPACK. By analyzing the conversations of teachers working in design teams, they have tracked the development of each of the seven components of TPACK over the course of a semester. This approach, however, is especially time-consuming and is methodologically specific to the unique context in which it was used (i.e., semester-long design experiences). Angeli and Valanides (2009) explored the use of design-based performance assessments embedded into course sequences. They used self-assessment, peer assessment, and expert assessment of these design-based performances as formative and summative assessments of teachers' understanding. They proposed five criteria for assessing TPACK in designing instruction with technology: (1) identification of topics to be taught with technology in ways that signify the added value of tools, such as topics that students cannot easily comprehend, or topics that teachers face difficulties in teaching effectively in class; (2) identification of representations for transforming the content to be taught into forms that are comprehensible to learners and difficultly supported by traditional means; (3) identification of teaching strategies, which are difficult or impossible to be implemented with traditional means; (4) selection of appropriate computer tools and effective pedagogical uses; and (5) identification of appropriate strategies to be combined with technology in the classroom, which includes any strategy that puts the learner at the center of the learning process.

Although Angeli & Valanides (2009) and Koehler et al. (2007) favored using the quantitative approach in data collection and analysis, the quantitative measures are relatively weak in portraying the subtle interrelationships among technology, pedagogy and content in the context of math teaching and learning. Furthermore, the quantitative measures are hardly used to assess and describe how teachers develop their TPACK. Therefore, in this study, we used Angeli and Valanides' (2009) five criteria for assessing the development of TPACK for the in-service math teacher. This study contained different types of qualitative data: the videos of teaching practices and community meetings, the transcripts and reflective journals about the interactions associated with teaching practices and the peer community group. Additionally, a questionnaire containing ten open-ended questions was designed for the purposes of conducting self-evaluation and peer-evaluation. The case teacher's development of TPACK was analyzed by using a triangular analysis with the data sources.

METHODOLOGY

This research employed the ideas of action research by Carr and Kemmis (1986), involving a spiral research process consisting of planning, acting, observing, reflecting and evaluating. To employ the process in the study, teachers revised instructional designs and plans through the process of reflections and changes, and through community group meetings and observations of students in order to reach the aims of the study.

Research Contexts

Participants

There were five elementary school teachers participating in the peer community group. However, the research focused on one of the teachers, Amy. She had been teaching students in lower grades for 12 years. In her 10th year of her teaching career, she was aware that she became less passionate in teaching as she had reached some degrees of familiarity with her curriculum based on her teaching experiences and spent less time and attention preparing teaching materials. With this awareness, she decided to apply for a graduate program in order to learn new pedagogical knowledge to become passionate in teaching again. Prior to this study, although Amy had experiences using IWBs in her math classes for one semester, she lacked sufficient skills in integrating the IWB into teaching for her TPACK development. Through the process of conducting the study, she expected to develop her profession in TPACK.

Background of the School and Teachers' Communities

The school where Amy taught provided clear objectives for the curriculum, and teachers were encouraged to form groups by classes for professional growth and support. Since June 2008, the school started installing an IWB to classrooms where teachers required one for their use. Most of the teachers only attended the professional workshops required by the school and seldom took the initiative to attend workshops for professional development. However, the workshops that required attendance by the school were mostly held as lectures. There was also a study group in the school, but the topics of the discussions varied and were rare in relation to teachers' professional development.

Student and Teaching Materials

There were 27 first-graders in Teacher Amy's class. Students who were first-time users of the IWB were curious and excited to try it when viewing the functions of IWBs such as animations, videos, pictures and games. During this study, one of the various math textbooks was used in class. This textbook included 9 units; the first circle contained the first 5 units: counting, length, location, order and quantity, plus and minus. The five units were all implemented with the use of the IWB. In the second circle, only unit 9, What time is it?, was instructed with the use of the IWB.

Study Design

The study contained two circles: the first circle was from August 2009 to November 2009; the second circle was from November 2009 to January 2010. Each circle was followed with the spiral process of planning, action, reflection and evaluation:

1) Planning: wrote the action research project and study plan for the community of teachers.
2) Action and observation: applied the IWB in teaching the mathematics units; interacted with the teachers in the peer community group.
3) Reflection: reflected on teaching practices and interactions of the community.

4) Evaluation: before the end of completing the two circles, the case study did a teaching demonstration. The case teacher self-evaluated her own teaching demonstration and was evaluated by other peers for the teaching and design through a TPACK questionnaire.

Data Collection and Analysis

The teaching content was based upon the mathematics textbooks designed for the first semester of first-grade. The community of teachers held meetings once every two weeks. The data sources were:

1) Reflective journals: the teacher's reflective journals of teaching; reflective journals after the meetings of the community.
2) Video-taped math lessons: math videotapes were watched and analyzed after classes.
3) Recordings of the meetings of the community: recordings were transcribed for data analyses.
4) TPACK questionnaire: the questionnaire was designed with open-ended questions based on the five criteria proposed by Angeli & Valanides (2009).

The study employed the steps of qualitative data analysis suggested by Patton (1990). The steps included: collecting raw data; organizing, classifying and coding the data, and analyzing the identified categories based on the research question and new findings classified in the categories. The collected data, according to different types of activities and timelines, was coded and classified in categories. After organizing and re-organizing the data, the results were finalized.

A Case Study: TPACK Development and Transformation

To better present the results, data sources, and time, researchers named teacher participants (except Amy) in the community as T1、T2、T3、T4 and students as S1, S2, and so on.

TPACK Development in the First Action Circle

Through "Learning by doing," Amy integrated the use of the IWB in her math teaching. In this circle, the teacher taught the five units by using the IWB: Counting, Length, Location, Orders, Plus and minus.

1. Other than evaluating a student's learning, the most important part in teaching is to evaluate teachers' own teaching problems.

   When we faced students' learning difficulties, we often tried to find out what problems students had and ignored the potential problems teachers might have in teaching. For example, teachers tended to ignore whether their expressions such as describing and posing the math problem was clear enough. Often times, mathematics teachers posed a math problem and then considered the rest of the part in solving the problem as mainly students' responsibilities, without considering if the problem was clearly stated.

T4: In the unit of Plus and Minus, students could concretely operate but did not understand what Minus and Plus meant.

T2: It's not his fault to divide 5 into 1+2+2. (Amy: No!)

T3: It's necessary to clearly explain how many portions are required for the number! (T1: It's indicated as 2 in the textbook)

Amy: It's teacher's problem when the problem is not clearly stated!

T2: The index in the textbook shows that it's possible to gradually pose the problem with several parts to help students understand the problem.

Community notes 20090820

2. Teachers should make math classes more flexible.

Teachers need to know the details of the content in math teaching. Merely following the materials attached to the textbook may take too much time in teaching the materials. After the discussions of the community, teachers found that Amy should not take the textbook as a superior guide; however, she needed to consider the actual conditions in class and adjust her own teaching strategies for students' cognitive development and understanding of representations. Teachers should evaluate appropriate teaching strategies and representations before teaching.

Amy: The materials attached to the textbook need to be used, and teachers often spend time allowing students to tear off the attached materials.

T4: I sometimes use other teaching tools as substitution such as replacing flower cards with real flowers!

T1: That's right! I didn't give them the flower cards! Sometimes I don't use the attached materials...

T3: Like I pasted a piece of flower onto a disc and asked them to show the operation on the stage. It is effective too!

Amy: It seems that I am inflexible!

Community notes 20090910

3. Cognition defines the role of IWBs in math classes

Although IWBs emphasize interactivity, it is impossible for all students to use an IWB simultaneously. The IWB is a tool for teachers to demonstrate and explain examples by providing various representations through teaching media. It is also a tool for individual students to demonstrate how they solve problems. IWBs are not appropriate for a whole class activity when requiring all students to participate the activity at the same time. Therefore, teachers need to find other ways to meet the needs. For example, additional documents from the textbook, teaching tools and small whiteboards are good for individual students to show their operation in solving the problem.

Before getting into the content, I first demonstrated the class activity "above and below" with the IWB to students. Students then took out the additional materials prepared ahead of time to join the activity. Students got involved during the process, and they could say the exact words for the locations above and below. For example, "There's a bell above the bear." Since the classroom is equipped with only one IWB set, only some students could demonstrate and explain their operation on the whiteboard. However, every student could learn through their operations by the demonstrations and explanations presented with IWBs and by individually operating the attached materials.

Reflective journal 20090923

4. Guides students to use the IWB sensor pen for their problem-solving.

Before allowing students to use the IWB, the teacher demonstrated its usage and functions to students. The first-grade students usually could not hold a pencil firmly and lacked writing experienc . It was difficult for them to hold the sensor pen to write well on the IWB such as writing with incorrect angles for holding a pen or holding a pen upside down. These difficulties resulted in the inability to write or drag with the sensor pen. To solve this problem, students practiced using the sensor pen and received immediate feedback and corrections from teachers during the class time. In addition, students who were interested in more practice could take turns to use the sensor pen when playing math games for interactivity from discs at recess time. Through all of these, students had chances to practice using the sensor pen with the functions of writing, dragging, and pointing in order to solve problems on the whiteboard.

S11: Teacher, how come the sensor pen is not working?
Amy: You're holding the senor pen upside down; the white buttons should be up!
S11: It's still not working!
Amy: Let me take a look….you turned off the light button!
Video 20090915

5. Effectively uses the IWB discs to identify the influence of IWBs on students' learning

With the applications of IWB in math teaching, the textbook sellers have already designed the integrated materials and related sources into the discs supplemented to the math textbook. What the teacher needed to do was to install the materials from the discs to a computer and flexibly use them in her math teaching.

(1) The animated videos stimulate students' motivation in learning mathematics.
The teacher could add math concepts related to the learning topics to the context of the story in the animated videos from the IWB discs. For students in lower grades, doing this could get their attention; therefore, the teacher played the animated video before each unit to enhance students' motivation.

Students had their first math class today. Some students had not gotten used to the school life yet so they were hesitant and anxious. When the students watched the animated video about seven little sheep, students suddenly paid attention to the screen. For first graders, animated images were attractive to them.

Reflective journal 20090902

(2) The electronic book helps students to stay on the learning subject and context. The first graders had short attention spans and often times lost attention to what page they were on and what activity they were involved in. The teacher could use the electronic book to provide hints to the students on the subject the teacher was teaching and through this way, students' learning could still continue if they forgot to bring their textbooks. In addition, first graders had an insufficient reading capability so the teacher needed to explain the meaning of each topic and demonstrate it. Teachers could save time in displaying contextualized pictures through the IWB instead of taking time to look for and flip over the pictures, to help students easily learn the topic.

When practicing, there was a way to combine the use of the electronic book and IWB to explain the meaning of the practices and demonstrate solutions to help students do the practices. From the perspectives of the evaluation, teachers combined useusing the electronic book and IWB to help students comprehend the meaning of their practices and reduce their misunderstandings. In this way, teachers could evaluate students' understanding of math concepts and leaning conditions.

Reflective journal 20090907

(3) IWB's interactive games promote student's motivation in learning, serving as a way of evaluation.
In the discs for the IWB, the setting of interactive games could get students' interest high enough that they had to answer correctly to move on to the next stage. Through playing the games, teachers could observe the process of how students played them to evaluate students' performance and understand students' comprehension of math concepts. Students' interests in learning were also facilitated.

The second activity in this unit was about testing the memory. Students remembered the locations of cards, covered them and then found out the paired cards. In the discs, the game was designed, and students only needed to use the sensor pen to play with it. Students all liked it, and most students could finish the equivalent amount of matching the cards. However, some students still needed more practice as they did not get used to interacting with the whiteboard with the sensor pen.

Reflective journal 20090909

1. Helps students' learning difficulties by using the IWB

During the community meetings, members often pointed out the math concepts students found difficult in learning after they taught the math concepts. Amy would first consider whether she could integrate the IBW into related teaching materials and transform the knowledge easily comprehensive to students. However, through the interactions among the community members, the researchers found that it would be more effective to provide demonstrations and operations for the concepts rather than using the IWB.

Amy: Students in my class had no difficulties in understanding some descriptive terms. However, when students did descriptive-based practices on paper, they felt confused with the descriptive terms of the concept.

T1: Some students did not know who should be the center in unit 3 "location".

T2: Because they did not understand the meaning of the problem.

T1: Through the text, students did not know who should be the center in deciding the locations of other people in the front and back.

Amy: Could teachers use IWB related sources to solve the difficulties?

T4: No, we should implement more contextualized teaching and provide more practices for students to construct the concept.

T2: I think we could ask students to form teams and ask them to tell the locations of their peers in the team.

Community notes 20091008

The Development of Case Teacher's TPACK during the First Research Circle

At the end of the first circle, Amy designed unit 5 "plus and minus" with detailed instructional procedures. In the community meetings, based on the teaching design and classroom videos, the members evaluated the case teacher's TPCK development in the first stage. Through the TPCK questionnaire of self-evaluation and peer-evaluation, Amy evaluated the qualitative data and described her TPCK development at the end of the first circle.

1. PCK development

   In the unit "plus and minus," the teacher identified students' misconceptions of abstract concepts and learning difficulties by asking students' questions, receiving feedback from students, and evaluating students' solutions to the problems on the whiteboard. To help students with their learning difficulties, the teacher could provide guidance and feedback based on students' different solutions towards obtaining the correct solution. Finally, the teacher gave an integrated explanation.

Amy: I gave more specific clarifications to different types of "minus" problems that students found easy to make mistakes.

T1: Amy could immediately give students feedback with explanations after students did the practice on the whiteboard.

T4: The teacher understood students' misconceptions by asking students questions and the solutions students gave on the whiteboard and then gave a solution of the problem.

T2: The teacher used the solutions students gave on small whiteboards to evaluate students' conceptual understanding of the content and provide immediate remedy when necessary.

T3: Amy provided more than two ways to solve "minus" problems that included exclusion and inclusion.

Circle 1 - Questionnaire 20091105

2. TPACK development

   An IWB was used as the main media for teachers to demonstrate explanations and for students to demonstrate their solutions of problems. For the whole class activity, students used small whiteboards to solve the problem in order for the teacher to provide immediate feedback and corrections whenever students found difficulties in solving the problem. In the meanwhile, students could learn from each other by presenting various ways in solving the problem. In this way, students lacked interactions and discussions among them since students solved the problem on their own. Interactions between the teacher and students were also insufficient.

Amy: We could see the interactions directing from the teacher to students, but lacked the interactions directing from students to the teacher. The interactions among students were even less.

T1: Students liked to solve the problem on the small whiteboards and even more interested in presenting their solutions to the class.

T4: Using the IWB to assist teaching and provide individual students with guidance to their solution of the problem on the small whiteboards not only helped students to complete the problem by themselves, but also revealed different ways of solving the problem. However, students lacked discussions.

T2: The teacher asked students questions and encouraged them to express, think, and share their ideas. However, questions from students were missing.

T3: Students were familiar with the teaching with the use of IWB and were encouraged to express their own ideas.

Circle 1 – Questionnaire 20091105

Conclusion of 1st circle

1) Based on the concept "Learning by doing," the case teacher understood the characteristics of the IWB and the role of IWBs played in the math course, such as the advantages of using IWB discs and the influences of IWBs on students' math learning.
2) The case teacher held a technology-centered pedagogical perspective when considering the teaching materials and instructional transformations and represent-tations. Amy looked for appropriate instructional strategies and representations centering on the use of IWBs, limiting her math teaching.
3) The teaching strategies lacked student-centered interactions and discussions. The results of the questionnaire showed that Amy used to teach with lectures; therefore, the interactions between students were insufficient.
4) This circle confirms the five stages: recognition, accepting, adapting, exploring and advancing of the TPACK development model proposed by Niess et al. (2009). Before the study, Amy had experienced the stages of recognizing, accepting and adapting. Amy was able to investigate the application of IWB sources and to recognize and define the role of IWBs in math class, showing her experiences entering the stage of exploring. Additionally, Amy was able to effectively use the discs of the IWB to define the influence of IWBs on students' math leaning, showing her teaching practices were approaching the advancing stage.

The points previously discussed provided the researchers with primary focuses in order to progress the study into the second circle. The focuses of the second circle for the case teacher were: (1) transforming technology-centered instruction to PCK-centered instruction based on the evaluations of using IWBs on various math topics, (2) accommodating and transforming the pedagogical strategies, and (3) enhancing the interactions and cooperation among students. These three points provided information for the case teacher to make changes in accommodating and transforming her current teaching strategies to new pedagogical approaches at the second circle.

Case Teacher's TPACK Development and Transformation during the Second Research Circle

1. Evaluates the proper use of IWB applications based on math topics

   After three months of students entering the elementary school, students were familiar with the schedule of class and were able to describe the order of events happening in their life. For misconceptions, some students could not distinguish the hour hand and the minute hand of the clock or ignore that the hour hand moves simultaneously with the minute hand. Other students were unable to predict the exact time of the morning or the afternoon and the daily schedule with reasonable time. Therefore, the IWB offered a large and clear fictional clock that the teacher could use the sensor pen to drag the minute hand so students could clearly see the change of the hour hand on the clock. Students learned the terminology of time (i.e., morning and afternoon) from their daily life experiences by using a video designed with students' concrete events happening in their daily lives. At the same time, students could retrieve back the sequence of the events and learn how to match the time and the events.

T3: I found that students in my class often dialed the hour hand and minute hand in opposite positions; they changed the time by dialing the hour hand to the wrong position instead of using the gear wheel in the back of the clock. For example, students directly dialed the hour hand to eight and the minute hand to six when I asked them to dial the clock to 8:30.

Amy: I told my students that they should use the gear wheel in the back of the clock to change the time instead of dialing the hour hand and minute hand directly.

T2: Teachers should emphasize that the short one is the hour hand, the long one is the minute hand and when the minute hand makes a complete rotation from 12 to 12, the hour hand only moves a small portion.

T4: Right, there were lots of students choosing "10PM" as their class time when they did their assignment.

Amy: This meant that students were not familiar with the terminology about time and real life experiences so they could not relate time to their daily life experiences.

T1: By using students' life events, we could ask them to verbally practice the terms in authentic situations.

Amy: In order to do this, a month ago, I started filming students' activities of daily life at school for every hour and 30 minutes (e.g., 8:30-look far, 11:00-clean classroom, 12:00-eat lunch, 12:30-line up to go home). We could teach this by simultaneously playing the videos in sequence, operating the clock, and using the terminology.

Community notes 20091231

2. Change of pedagogy and teaching strategy – students can make math problems

   Students could make math problems and try to solve the problem at the same time. Since they were the ones to make the problem and find out the solution of the problem, they were involved in doing this.

In the unit "What time is it?" there're usually activities from the textbook that call for teachers telling the time and students dialing a clock for the respective time. Since the first graders were capable of describing their daily life schedules, I asked them to make math problems according to their real life and had other classmates take a guess at what events could correspond with the time as well as practicing the terms of morning and afternoon. Learning became interesting when it was related to students' daily life.

Reflective Journal 20091230

3. Enhances the cooperative and interactive activities among students:

   Arranges students' seats and guides their discussions in groups

   Through the feedback from other teachers and changing her own beliefs, Amy believed that it would be beneficial for students to learn through interactions, cooperation and observations among students. She rearranged students' seats with four or five students sitting face to face in a group, benefiting the discussions and observations among students.

Amy: I need to work harder at interacting with my students, but I do not know where I should start?

T1: You can start with rearranging their seats. You need to divide them into groups, so they can have group discussions.

T2: You better expect that the class may be noisy.

T4: It should be fine in Amy's class. Her students are well-behaved.

Amy: I will tell them the rules and ask them to follow.

T1: You could give a try first and make adjustments if necessary.

T3: Students may not know how to discuss with others so you need to assist them.

Amy: I will join them depending on the situations and use questioning to guide them.

T1: You have thought about it. So just do it.

Community notes 20091119

The Case Teacher's TPACK Development in the Second Stage

1. PCK development

   The math concept *Time* is a difficult topic for students to comprehend and easily causes misconceptions. For example, students might get confused with the minute hand and hour hand of a clock; they might ignore the simultaneous movement of the minute hand and hour hand. Other than using the IWB to explain and demonstrate the concept, every student played with the clock and observed different locations of the hour and minute hand. Amy could also give students immediate feedback and remind them not to dial the hour hand by hand.

Amy: Through operating the clock, students could observe the differences between the hour hand and minute hand in the class.

T4: The teacher already considered students' confusion with the relations of the hour and minute hands in designing the lesson. Students could make observations to realize the relations through operating the clock.

T2: For those students who made mistakes in operating the clock, teachers could give immediate feedback and make corrections.

T3: The teacher used a fictitious clock to clearly define the meaning of the minute hand going a half round to show the differences between morning and afternoon.

T1: The teacher repeatedly reminded the students not to dial the hour hand by hand; they should use the gear wheel in the back of the clock.

Circle 2 Questionnaire 20091231

2. TPACK development

With the features of the IWB, the teacher used IWBs fordemonstrations, and displaying solutions and students' presentations. Considering the student-centered perspectives, Amy had students make math problems in groups so they could interact, discuss and collaborate with each other. Group members tried to solve the problem while making problems to check whether the solution from other groups was correct, at the same time, students' ability of solving problems collaboratively was enhanced.

Amy: The videos made with students' daily schedule related to their life. Students collaboratively made and solved the problems in a group at the same time.

T4: The teacher used the IWB to allow students to present the solution and other students to check whether the solution was correct as well as practice critical thinking. The teacher gave feedback when needed and students made observations.

T2: The teacher encouraged students to design problems in groups and give operations with the IWB so other students could observe and learn.

T3: The teaching design was student-centered to encourage students to express and solve the problem collaboratively.

T1: The integrated activities allowed students to design problems together, which could enhance students' cooperation, problem solving and their learning interest.

Circle 2 Questionnaire 20091231

Based on the feedback received from circle 1, Amy's teaching style transformed from Technology-centered to a PCK-centered mode. Considering the math content and students' characteristics, Amy evaluated the application of technology and used the IWB as a crucial tool to transform teaching materials and integrate diversified sources. The IWB helped teachers transform the difficult and abstract concepts into the forms comprehensible to students. In circle 2, the case teacher adjusted her pedagogical strategies with the student-centered perspectives. Therefore, Amy made videos based on students' real life to help students learn concepts associated with time and the clock. Teachers with TPACK need to design classroom activities in various forms, including lectures and hands-on activities (Angeli & Valanides, 2009). To improve students' cooperation and interactions, Amy started rearranging students to sit face-to-face so students could discuss with and observe from each other. Students in their own group made math problems collaboratively, and their interactions were enhanced. In circle 2, the teacher's teaching strategies became diverse, compared to the teacher-centered strategies in circle 1.

Discussion and Implications

From the prior literature review, the researchers who proposed the concept of TPACK (Koehler & Mishra, 2005) and the TPACK development model for math teachers (Niess et al., 2009) mainly had a background of information technology. When discussing the PCK in the TPACK model, they mentioned the knowledge of the subject content and students' prior knowledge, but the in-depth discussions were missing. Although the model includes Technology, Pedagogy and Content, they often over emphasize technology and ignore PCK in teaching.

Koehler and Mishra (2005) proposed the concept of TPCK and discussed the relationships of three knowledge elements: Technology, Pedagogy and Content. However, current research lacks the discussions on the knowledge related to students and subject content knowledge, clearly defining and explains the integration of technology. The discussions place primary focus on T instead of P and C. The TPACK development model proposed by Niess et al. (2009) viewed teaching particularly with the emphasis on technological perspectives, lacking student-centered strategies and limiting teachers' teaching. In the first circle, since Amy was devoted in using IWB in math class, she held IWB-based pedagogy. Thus, interactions and collaborations among students were insufficient.

In summary, the researchers did an action research by employing the TPACK framework. In the first circle, the case teacher mainly focused on the use of IWBs the same as the results from prior studies. After going through the circles, she adjusted and reoriented her pedagogical strategies towards student- and content-centered strategies, and considered technology as a tool to supplement teaching. Nowadays, the use of technology is becoming essential and teachers' TPACK development is important. When emphasizing the development of TPACK, PCK should also be considered as an important factor with the integration of technology.

The study discussed the case study teacher's TPACK development and transformation. She had more than ten years of teaching experience as she had the knowledge of teaching materials, and students' prior knowledge and backgrounds. In adding the new element of technology to PCK, the teacher's TPACK was developed. Angeli & Valanides (2009) indicated that the most significant difference between experienced teachers and novice teachers was their knowledge about their students. Novice teachers just begin their teaching career so their TPACK development would be different from experienced teachers. Further research is needed to examine the transformations with the use of IWBs that may occur in novice teachers.

After a semester with the interactions in the community, the results showed that the community group was beneficial for teachers' professional development. Other than experienced teachers, an effective community should also include teachers with subject and technology expertise. This study included five teachers, and one of them was a teacher with math expertise. Although a teacher with technology expertise was included in the community group, he could not regularly attend the meetings and discussions. If teachers with subject and technology expertise could brainstorm the pedagogical strategies integrated with technology with the focus of knowledge on subject content and students, there would be more pedagogical innovations.

Prior research suggested that teachers develop their TPACK by defining the use of technological media to develop their TPACK (Dalgarni & Colgan, 2007; Koehler & Mishra, 2005; Koehler et al., 2007). Thus, IWBs in the study were designed as a technological media to develop the teacher's TPACK. From the feedback of the circles, it was easy for the teacher to focus on technology-centered pedagogy and limit the transformations of the content that might occur. The study suggested that future researchers overview various forms of information and technology integration to better examine teachers' overall TPACK development.

REFERENCES

Angeli, C. & Valanides, N. (2009). Epistemological and methodological issues for the conceptualization, development, and assessment of ICT–TPCK: Advances in technological pedagogical content knowledge (TPCK). *Computers & Education, 52,* 154–168.

Bowman, C. & McCormick, S. (2000) Comparison of peer coaching versus traditional supervision effects, *The Journal of Educational Research, 93(4),* 256-264.

Carr, W. & Kemmis, S. (1986). Becoming critical: Education, knowledge and action research. Basingstoke: Falmer Press.

Dalgarno, N. & Colgan, L. (2007). Supporting novice elementary mathematics teachers' induction in professional communities and providing innovative forms of pedagogical content knowledge development through information and communication technology. *Teaching and Teacher Education, 23,* 1051-1065.

Gillen, J., Littleton, K., Twiner, A., Staarman, J. & Mercer, N. (2008). Using the interactive whiteboard to resource continuity and support multimodal teaching in a primary science classroom. *Journal of Computer Assisted Learning, 24,* 348–358.

Glover, D., Miller, D., Averis, D. & Door, V. (2005). The interactive whiteboard: A literature survey. *Technology, Pedagogy and Education, 14(2),* 155-170.

Glover, D., Miller, D., Averis, D. & Door, V. (2007). The evolution of an effective pedagogy for teachers using the interactive whiteboard and modern languages: An empirical analysis from the secondary sectors. *Learning, Media and Technology, 32(1),* 5-20.

Grossman, P. L. (1990). The making of teacher: Teacher knowledge and teacher education. New York: Teacher College Press.

Hall, I., & Higgins, S. (2005). Primary school students' perceptions of interactive whiteboards. *Journal of Computer Assisted Learning, 21,* 102–117.

Hennessy, S., Deaney, R., Ruthven, K. & Winterbottom, M. (2007). Pedagogical strategies for using the interactive whiteboard to foster learner participation in school science. *Learning, Media and Technology, 32,* 283–301.

Higgins, S., Beauchamp, G. & Miller, D. (2007). Reviewing the literature on interactive whiteboards. *Learning, Media and Technology, 32(3),* 213-225.

Holmes, K. (2009). Planning to teach with digital tools: Introducing the interactive whiteboard to pre-service secondary mathematics teachers. *Australasian Journal of Educational Technology, 25(3),* 351-365.

Jang, S. J. (2010). Integrating the interactive whiteboard and peer coaching to develop the TPACK of secondary science teachers. *Computers & Education*, 55(4), 1744-1751.

Jenkins, J. & Veal, M.L. (2002). Preservice teachers' PCK development during peer coaching. *Journal of Teaching in Physical Education, 22(1),* 49-68.

Jenkins, J. Garn, A. & Jenkins, P. (2005). Preservice teacher observations in peer coaching. *Journal of Teaching in Physical Education, 24*, 2-23.

Kennewell, S. (2001) Using affordances and constraints to evaluate the use of ICT in teaching and learning, *Journal of IT and Teacher Education, 10,* 101-116.

Koehler, M.J. & Mishra, P. (2005). What happens when teachers design educational technology? The development of technological pedagogical content knowledge. *Journal of Educational Computing Research, 32(2),* 131-152.

Koehler, M.J., Mishra, P. & Yahya, K. (2007). Tracing the development of teacher knowledge in a design seminar: Integrating content, pedagogy and technology. *Computers & Education, 49,* 740–762.

Lave, J. & Wenger, E. (1991). Situated learning: Legitimate peripheral participation. Cambridge, UK: Cambridge University Publishers.

Lopez, O. S. (2010). The digital learning classroom: Improving English language learners' academic success in mathematics and reading using interactive whiteboard technology. *Computers & Education, 54,* 901–915.

McAllister, E. A. & Neubert, G. A., (1995). New teachers helping new teachers: Preservice peer coaching. Bloomington, In: EDINFO Press.

Miller, D. & Glover, D. (2002). The interactive whiteboard as a force for pedagogic change: The experience of five elementary schools in an English education authority. *Information Technology in Childhood Education Annual, 2002(1)*, 5-19.

Miller, D., Glover, D. & Averis, D. (2005). Developing pedagogic skills for the use of the interactive whiteboard in mathematics. *British Educational Research Association, Glamorgan.* Retrieved from http://bit.ly/iwbpedagogy05bera

Niess, M. L. (2005). Preparing teachers to teach science and mathematics with technology: Developing a technology pedagogical content knowledge. *Teaching and Teacher Education, 21(5),* 509–523.

Niess, M. L., Ronau, R. N., Shafer, K. G., Driskell, S. O., Harper S. R., Johnston, C., Browning, C., Özgün-Koca, S. A. & Kersaint, G. (2009). Mathematics teacher TPACK standards and development model. *Contemporary Issues in Technology and Teacher Education, 9(1),* 4-24.

Patton, M. Q. (1990). *Qualitative evaluation and research methods (2nd.)*. London: Sage.

Pierce, D. & Hunsaker, T. W. (1996) Professional development for the teacher, of the teacher, and by the teacher, *Education, 117(1),* 101-105.

Schmid, E. C. (2008). Potential pedagogical benefits and drawbacks of multimedia use in the English language classroom equipped with interactive whiteboard technology. *Computers & Education, 51,* 1553–1568.

Sedig, K. & Liang, H. (2006). Interactivity of visual mathematical representations: Factors affecting learning and the cognitive process. *Journal of Interactive Learning Research, 17(2),* 179-212.

Shenton, A. & Pagett, L. (2007). From 'bored' to screen: The use of the interactive whiteboard for literacy in six primary classrooms. *Literacy, 41(3),* 129-136.

Shulman, L. S. (1987). Knowledge and teaching: Foundations of the new reform. *Harvard Educational Review, 57,* 1-22.

Smith, F., Hardman, F. & Higgins, S. (2006). Impact of interactive whiteboards on teacher–pupil interaction in the national literacy and numeracy strategies. *British Educational Research Journal, 32(3),* 443–457.

Smith, H. J., Higgins, S., Wall, K. & Miller, J. (2005). Interactive whiteboards: Boon or bandwagon? A critical review of the literature. *Journal of Computer Assisted Learning, 21*, 91-101.

Taylor, M., Harlow, A. & Forret, M. (2010). Using a computer programming environment and an interactive whiteboard to investigate some mathematical thinking. *Procedia Social and Behavioral Sciences, 8, 561–570*

Torff, B. & Tirotta, R. (2010). Interactive whiteboards produce small gains in elementary students' self-reported motivation in mathematics. *Computers & Education,* 54, 379-383.

Walker, D.(2003). Quality at the Dockside. TES Online, 2, 66 -67.

Warwick, P. & Kershner, R. (2008). Primary teachers' understanding of the interactive whiteboard as a tool for children's collaborative learning and knowledge-building. *Learning, Media and Technology, 33(4),* 269-287.

Wood, D. (2001). In search of fluency: What is it and how can we teach it? *Canadian Modern Language Review, 57* (4), 573-589.

INDEX

A

B

C

G

H

I

J

K

L

M

R

S

T